MASTERPIECES

BRIDGE
ARCHITECTURE + DESIGN

M A S T E R P I E C E S

CHRIS VAN UFFELEN

BRIDGE
ARCHITECTURE + DESIGN

BRAUN

PREFACE. VORWORT. PRÉFACE.

A long time ago a tree fell across a brook and became a bridge. Our ancestors saw this, copied it and bridges were among the first constructions made by mankind. Soon the deck was made of stone slabs and at one point a central pillar was added to bridge longer distances (Postbridge, Dartmoor). The first suspension bridges evolved in Asia and Africa in the Bronze Age already, and arched bridges were invented in Mesopotamia. The arched construction style allowed the Romans to perfect the art of building stone bridges (Pont du Gard, Nîmes, 14 A.D.). The first iron chain bridges were probably already erected in China in the 3rd century AD.

During the Middle Ages the ancient bridges continued to be in use, but at the same time important bridges were erected such as the Pont d'Avignon (1188) and the London Bridge (1209), frequently with bridge towers, such as the Charles Bridge in Prague (from 1357). These portals highlight the threshold character of the constructions, while other additions, such as bridge chapels or the use as a shopping street (Ponte vecchio, Florence, 1345) emphasize the connection character. While such commercial use was common in Paris, the oldest preserved bridge, the Pont Neuf (1607) was the first bridge without such constructions, to keep

Brücken existieren länger als die Menschheit. Irgendwann in grauer Vorzeit kippte ein Baum über einen Bach und wurde zur Brücke. Unsere Urahnen sahen es, machten es nach und Brücken wurden zu den ersten von Menschenhand geschaffenen Bauwerken. Bald wurde das Deck aus einer Steinplatte errichtet und irgendwann kam ein Mittelpfeiler zur Überbrückung größerer Distanzen hinzu (Postbridge, Dartmoor). Erste Hängebrücken entstanden in Asien und Afrika schon in der Bronzezeit und in Mesopotamien wurde die Bogenbrücke erfunden. Durch den Bogenbau perfektionierten die Römer den steinernen Brückenbau (Pont du Gard, Nîmes, 14 n. Chr.). Erste eiserne Kettenbrücken entstanden in China wohl schon im 3. Jh. n. Chr.

Im Mittelalter wurden die antiken Brücken weiter genutzt, aber es entstanden auch bedeutende Brücken wie die Pont d'Avignon (1188) und die London Bridge (1209), häufig mit Brückentürmen, wie bei der Prager Karlsbrücke (ab 1357) durch Peter Parler ergänzt. Diese Portale verstärken den Schwellcharakter der Bauten, während andere Anbauten, so die Brückenkapellen oder die Nutzung als Einkaufsstraße (Ponte vecchio, Florenz, 1345) den Verbindungscharakter betonen. Diese gewerbliche Nutzung gab es in Paris allenthalben, doch ist die hier älteste erhaltene Brücke, die Pont Neuf (1607), ausgerechnet die erste ohne solche Aufbauten, so dass der Blick auf den

Les ponts sont antérieurs à la civilisation : ayant vu un arbre tomber en travers d'un ruisseau, un homme de la Préhistoire a imité la nature, de sorte que le pont est devenu le premier ouvrage d'art de l'histoire de l'Humanité. Les premiers ponts en pierres ont bientôt été complétés par une pile centrale permettant de franchir une plus grande distance (Postbridge, Dartmoor). Des ponts suspendus sont apparus en Asie et en Afrique dès l'âge du Bronze, tandis que les Mésopotamiens inventaient le pont voûté, la technique étant perfectionnée par les Romains (pont du Gard, 14 apr. J.-C.). Les Chinois ont quant à eux construit des ponts caténaires métalliques dès le IIIe siècle de notre ère.

Au Moyen Âge, les ponts antiques, toujours utilisés, se complètent de nouveaux ouvrages comme le pont d'Avignon (1188) ou le pont de Londres (1209), souvent protégés par des tours telles que celles du pont Saint-Charles de Prague (1357). Ces éléments défensifs soulignent la dimension frontalière des ponts, tandis que d'autres ajouts, notamment des chapelles ou des boutiques (Ponte Vecchio, Florence, 1345) mettent en valeur la communication et les échanges. Les échoppes se généralisent progressivement, au point que le Pont Neuf (1607, le plus vieux des ponts parisiens) inverse la tendance puisqu'il est le premier à ne pas porter de boutiques

the view of the Louvre intact. It became the model for further development, which, with the increased physical knowledge in the age of reason, resulted in the École des Ponts et Chaussées (1747), the world's first engineering school. The constructions of its teacher Jean Perronet, "father of modern bridge building" were seen as an affront to academic architecture.

At the same time, iron was developed as building material (Coalbrookdale cast iron bridge, 1779). In the 19th century, bridge building became professional through the influence of engineering schools and public officials (such as Telford and Schinkel). Many bridges were built for the new railways (Britannia bridge, Menai street, 1850), Gustave Eiffel perfected systems building (Douro bridge, Porto, 1877), vastly expanding the dimensions and culminating in the Firth of Forth Bridge (Scotland, 1889) and the Brooklyn Bridge (New York, 1883.

The advent of steel in 1890, which replaced the iron that had been used since 1840, allowed the construction of bold suspension bridges (Golden Gate Bridge, San Francisco, 1937). Since the early 20th century, reinforced concrete was increasingly used (François Hennebique: 1894 in Wiggen). Robert Maillart and Eugène Freyssinet developed the suitable shapes for this material. At this time, bridges had long since started to cope with mass transit and had a higher carrying capacity that was frequently also displayed (220 Amsterdam bridges P. L. Kramers, 1917-28).

By the mid-20th century, pre-stressed concrete was primarily used in bridge construction. Since that time, the ongoing race for faster, higher and further led to new problems and solutions. In contrast to the focus on functional aspects of bridge architecture in late modernism today's bridge constructions display a greater extent of esthetic value.

Louvre unverstellt blieb. Sie wurde Vorbild für die weitere Entwicklung, die im Zeitalter der Vernunft durch zunehmende physikalische Einsichten zur École des Ponts et Chaussées (1747) als erster Ingenieursschule der Welt führte. Die Bauten ihres Lehrers Jean Perronet, „Vater des modernen Brückenbaus", wurden als Affront gegen die akademische Baukunst empfunden.

Zeitgleich wurde auch Eisen als Baumaterial erschlossen (Gußeisenbrücke Coalbrookdale, 1779). Im 19. Jahrhundert professionalisierte sich der Brückenbau durch den Einfluss von Ingenieursschulen und Amtsträgern (Bauinspektor Telford; Oberbaurat Schinkel). Zahlreiche Brücken entstanden für die neuen Eisenbahnen (Britannia-Brücke, Menai-Straße, 1850), Gustave Eiffel perfektionierte den Systembau (Douro-Brücke, Porto, 1877) und die Dimensionen steigerten sich enorm. Die Firth of Forth Bridge (Schottland, 1889) sowie die Brocklyn Bridge (New York, 1883), stellten einen vorläufigen Höhepunkt dar.

Erst der Einsatz von Stahl, der das seit 1840 beherrschende Gusseisen um 1890 ablöste, machte kühne Hängebrücken möglich (Golden Gate Bridge, San Francisco, 1937). Seit Anfang des 20. Jahrhunderts wurde aber auch Stahlbeton (François Hennebique: 1894 in Wiggen) zunehmend genutzt. Hier fanden Robert Maillart und Eugène Freyssinet die dem Material angemessenen Formen. Zu dieser Zeit hatten die Brücken längst mit dem Massenverkehr zu kämpfen und eine größere Tragkraft zu haben, die oft auch gezeigt wurde (220 Amsterdamer Brücken P. L. Kramers, 1917-28).

Zur Mitte des 20. Jahrhunderts übernahm der Spannbeton die Vorherrschaft im Brückenbau. Seit jener Zeit steigerte sich das schneller, höher und weiter ununterbrochen und es ergaben sich neue Probleme und Lösungen. Die Grundprinzipien und Aufgaben blieben hingegen gleich. Nach dem funktionalen Brückenbau der Spätmoderne ist die Aufgabe heute längst wieder vorrangig eine ästhetische.

de manière à offrir une perspective sur le Louvre. Après la fondation de l'École des ponts et chaussées (1747, la plus ancienne école d'ingénieurs au monde), Jean Perronet, considéré comme le « père des ponts modernes », réalise des ouvrages qui bouleversent les règles académiques de l'époque.

La fonte fait peu après son apparition (Coalbrookdale, 1779) et au XIXe siècle, la construction des ponts se professionnalise sous l'influence d'ingénieurs tels que Thomas Telford et de grands architectes tels que Karl Friedrich Schinkel. De nombreux ponts de chemin de fer doivent alors être construits (pont du détroit de Menai, 1850), ce qui incite Gustave Eiffel à perfectionner la technologie des ouvrages préfabriqués (pont sur le Douro, 1877). Parallèlement, on assiste à un accroissement considérable des distances franchies, le pont de Brooklyn (New York, 1883) et celui sur le Firth of Forth (Écosse, 1889) établissant des records.

À partir de 1890, le remplacement de la fonte par l'acier permet de réaliser d'audacieux ponts suspendus (Golden Gate Bridge, San Francisco, 1937), tandis que l'emploi du béton armé se généralise sous l'influence de François Hennebique (pont de Wiggen, 1894). Au début du XXe siècle, les ponts doivent faire face à l'augmentation de la circulation et des masses transportées, ce qui donne lieu à un regain de construction. C'est ainsi qu'entre 1917 et 1928, Pieter Lodewijk Kramer construit à lui seul plusieurs centaines de ponts.

Dans les années 1950, la généralisation du béton précontraint permet de construire toujours plus rapidement des ouvrages toujours plus hauts et plus longs. Après les ponts strictement fonctionnels des décennies précédentes, on en revient maintenant à des ouvrages où l'esthétique joue un rôle accru, le présent ouvrage offrant un aperçu de la variété des solutions imaginées par les architectes contemporains.

PROJECTS. PRO

JEKTE. PROJETS.

WEBB BRIDGE,
MELBOURNE, VIC, AUSTRALIA

DENTON CORKER MARSHALL WITH ROBERT OWEN

www.dentoncorkermarshall.com

Type: Box beam bridge, **Length:** 225 m, **Artist:** Robert Owen, **Structural and Lighting engineers:** ARUP, **Completion:** 2003, **Photos:** Shannon McGrath.

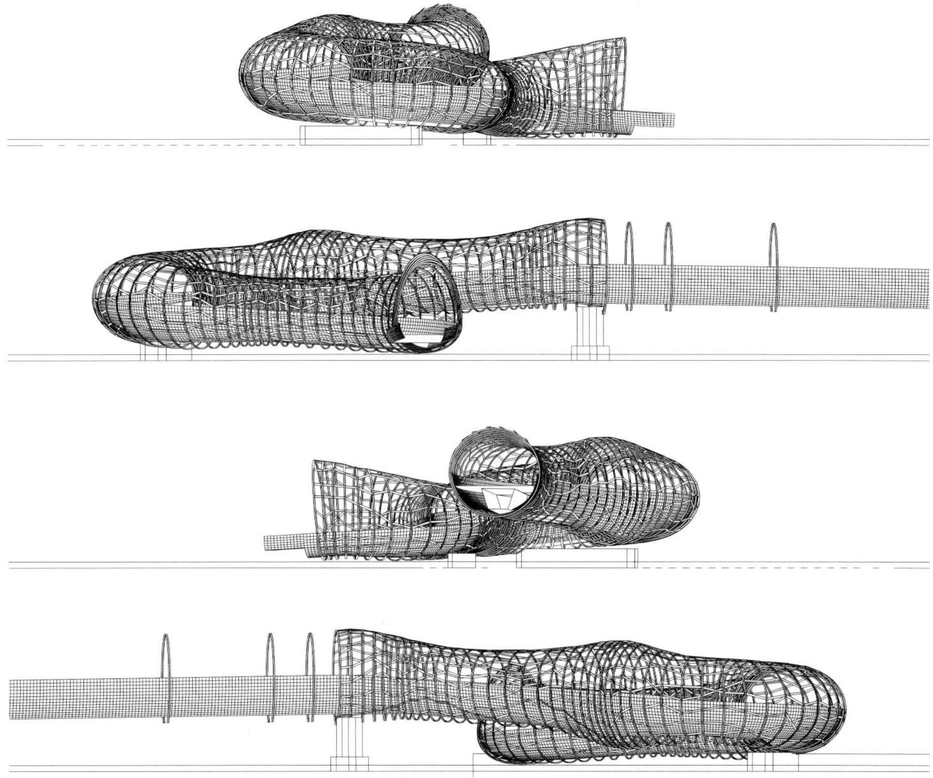

Left: Box beam. Links: Hohlkasten. À gauche: Superstructure. | Right: Elevations. Rechts: Aufrisse. À droite: Vues en élévation.

The bridge had to re-use remaining sections of the old rail bridge to link the Docklands to a new residential development. The existing structure (145 meters) and the new curved connecting link (80 meters), which takes up level changes and creates a point of arrival at the south bank, are joined seamlessly in a voluminous pliant and curved shape. The unified sculptural form consists of a pigmented concrete screed deck on steel box beams and the containing baskets/ribs of circular/oval hoops that encircle the bridge deck. The hoops vary in width from 5 to 8.7 meters and in height from 4 to 8.9 meters at varying centers along the bridge.

Die Brücke musste verbliebene Teile der alten Gleisbrücke wiederverwenden, um das Hafenviertel mit einem Wohngebiet zu verbinden. Der überlieferte (145 Meter) und der neue, sich windende, und Niveauunterschiede ausgleichende Bauteil (80 Meter), der ein Foyer für das Südufer bildet, sind nahtlos zu einer voluminösen, geschneidig und gebogenen Form verbunden. Die vereinheitlichte skulpturale Form besteht aus dem Deck aus pigmentiertem Betonestrich auf stählernen Hohlkästen und dem Flechtwerk/Rippen kreisförmiger/ovaler Reifen, die das Deck umschließen. Sie entsteht durch Variation der Reifengrößen (Breite 5 bis 8,7, Höhe 4 bis 8,9 Meter) und -mittelpunkte.

Une passerelle relie les docks de Melbourne à un nouveau quartier résidentiel. Elle se compose d'un ancien pont de chemin de fer de 145 mètres de long et d'une structure moderne de 80 mètres qui compense les différences de hauteur sur la rive gauche. Cet ensemble homogène aux formes incurvées et sculpturales est pourvu d'un tablier en ciment pigmenté reposant sur des caissons en acier, ainsi que d'une superstructure au profil tantôt ovale, tantôt circulaire dont les éléments verticaux évoquent des côtes ou des anses de paniers. La largeur de la superstructure varie de cinq à 8,70 mètres, sa hauteur de quatre à 8,90 mètres.

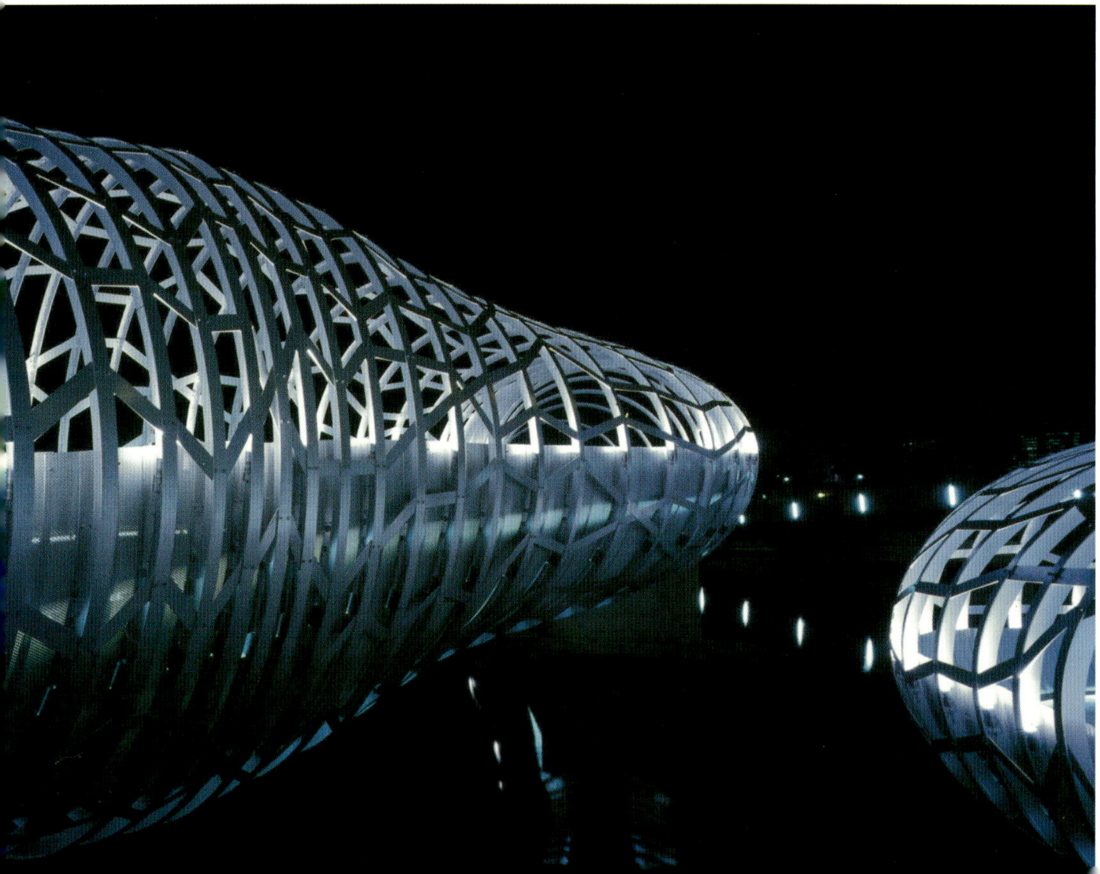

From left to right, from above to below:
Old and new part, in the loop, entrance.
Right: Loop.

Von links nach rechts, von oben nach unten:
Alter und neuer Teil, in der Schlaufe, Eingang.
Rechts: Schlaufe.

De gauche à droite, du haut vers le bas:
Ancien pont et nouvelle structure,
vue de nuit, entrée de la passerelle.
À droite: Vue de la boucle.

CRAIGIEBURN BYPASS,
MELBOURNE, VIC, AUSTRALIA

TONKIN ZULAIKHA GREER, TAYLOR CULLITY LETHLEAN

www.tzg.com.au, www.tcl.net.au

Type: Curved box truss, **Length:** 65 m, **Structural engineers:** Meinhardt, **Artist:** Robert Owen, **Completion:** 2005, **Photos:** Peter Hyatt (14, 16 b. l., 17), John Gollings (16 a., 16 b. r.).

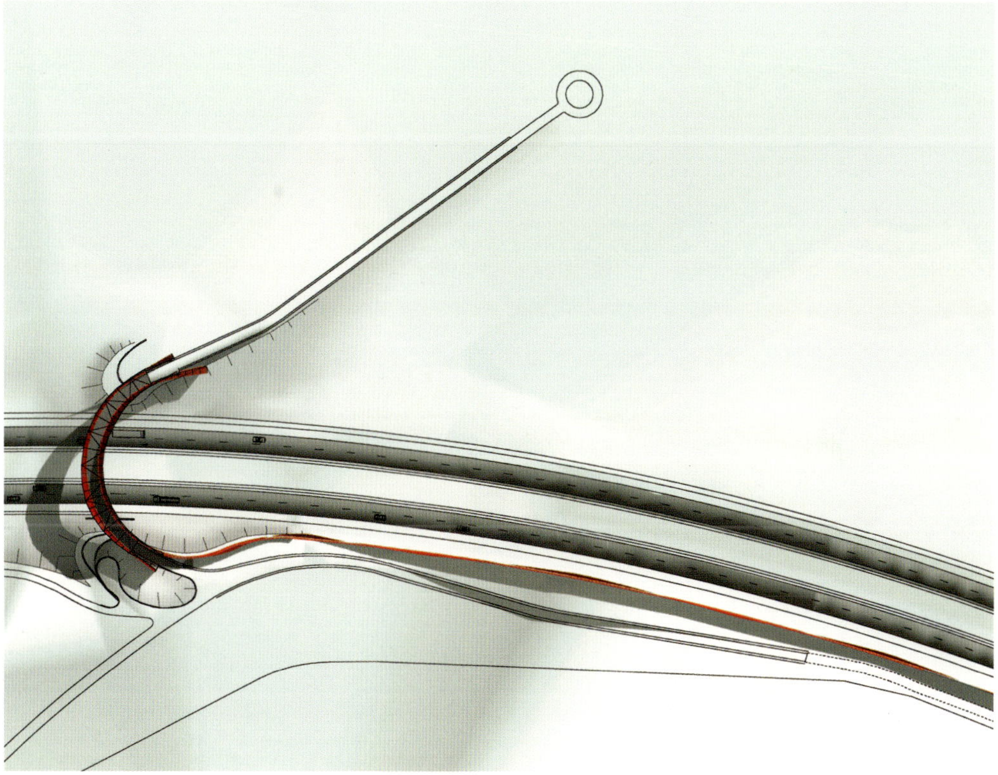

Left: Pedestrian bridge. Links: Fußgängerbrücke. À gauche: Passerelle. | Right: Site plan. Rechts: Übersichtsplan. À droite: Plan de situation.

This project along a 32-kilometer stretch of freeway linking the Hume Highway with the Melbourne Ring Road was designed to be experienced at a regular freeway speed. It includes three series of sculptural sound walls, a pedestrian bridge, and a set of design parameters for road bridges, crash barriers and retaining structures. A major element of the work was the new pedestrian bridge, which was designed as a gateway to the distant city of Melbourne, visible on the horizon. The bridge, with its complex plan and elevation curves, is a tubular steel truss covered in the same austenitic steel as the main sound walls, which at this point appear to leap over the road in a gesture of welcome or farewell.

Dieses Projekt entlang eines 32 Kilometer langen Streifens Autobahn, welche den Hume Highway mit der Melbourne Ring Road verbindet, wurde so gestaltet, dass es bei Reisegeschwindigkeit wahrgenommen werden kann. Es besteht aus drei Reihen skulpturaler Lärmschutzwände, einer Fußgängerbrücke und einer Gruppe von Straßenbrücken und Leitplanken. Eine neue Fußgängerbrücke, die als Tor nach Melbourne gestaltet wurde, bildet das Hauptelement. Diese als röhrenförmiger Stahlträger ausgebildete Brücke formt in Grundriss und Aufriss einen sphärischen Bogen. Sie ist mit demselben austenitischen Stahl verkleidet wie die Schallschutzmauern, die hier über der Straße hervorspringen als wollten sie den Reisenden grüßen.

Cette passerelle qui franchit un tronçon d'autoroute de trente-deux kilomètres de long entre le périphérique de Melbourne et l'autoroute de Hume a été conçue pour être appréciée en roulant à la vitesse maximale autorisée. Il s'agit d'un ensemble composé de trois murs antibruit, de la passerelle proprement dite, de glissières de sécurité et de structures de retenue. Du haut de la passerelle, on aperçoit la ville de Melbourne à l'horizon. Cette structure tubulaire en acier, incurvée horizontalement et verticalement, est garnie sur les côtés de plaques d'acier Corten similaires à celles des murs antibruit et légèrement inclinées vers l'autoroute comme dans un geste de bienvenue ou d'adieu.

From left to right, from above to below: Pedestrian bridge, aerial view, curtain wall and pedestrian bridge with view of Melbourne. Right: Inside the pedestrian bridge.

Von links nach rechts, von oben nach unten: Fußgängerbrücke, Luftaufnahme, Wand und Fußgängerbrücke Blickrichtung Melbourne. Rechts: Im Inneren der Fußgängerbrücke.

De gauche à droite, du haut vers le bas: Passerelle vue de l'autoroute, vue aérienne, mur antibruit avec la passerelle et Melbourne à l'arrière plan. À droite: Vue du tablier.

BRICK PIT RING,
SYDNEY, NSW, AUSTRALIA

DURBACH BLOCK ARCHITECTS

www.durbachblock.com

Type: Deck truss with trestle support, **Length:** 552 m, **Planning partners:** Sydney Olympic Park Authority, **Client:** Sydney Olympic Park Authority, **Completion:** 2005, **Photos:** Kraig Carlstrom (18, 20 a. l., 20 a. r.), Peter Hyatt (20 b., 21).

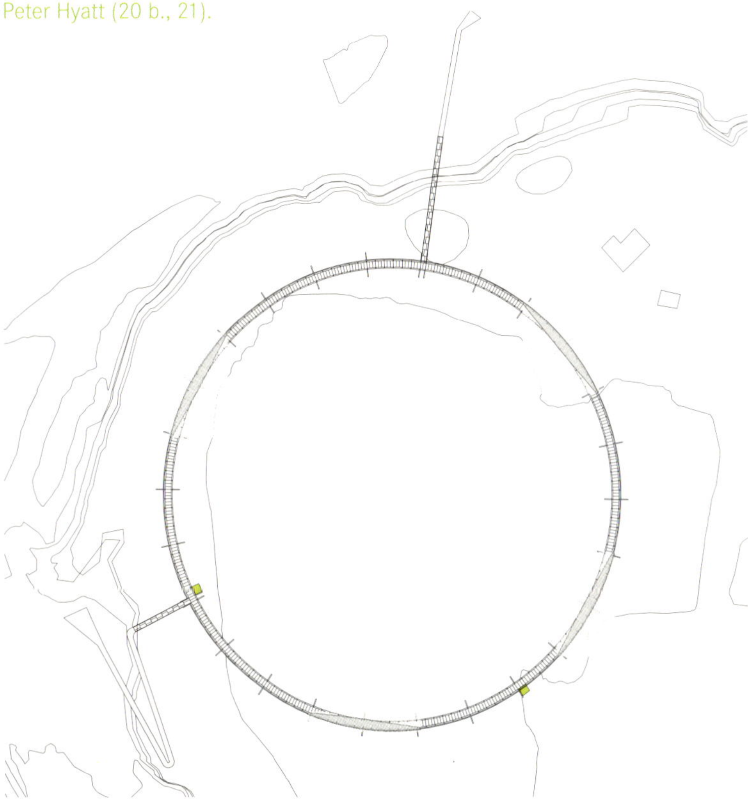

Left: Overall view. Links: Gesamtansicht. À gauche: Vue d'ensemble. | Right: Floor plan. Rechts: Grundriss. À droite: Plan.

Brick Pit Ring located near Sydney is an environmental information center and outdoor exhibition. It reconciles public access with safety and wildlife conservation in a manner that transforms these ambitions into an iconic structure. The transparent building is beautiful in its simplicity and poetic in its vision. The structure consists of a tour path resting on steel piles. The path is framed with a parapet that provides viewing slots here and then. The style of the structure is exemplary for constructions in fragile habitats and deliberately explores three themes: contrast without contradiction, access without contact, and monumentality without weight.

Der Brick Pit Ring in der Nähe von Sydney ist zugleich Umweltinformationszentrum und Ausstellungsanlage. Es vereint öffentlichen Zugang mit Sicherheit und Schutz von Wildtieren in einer Weise, dass diese Ziele zu einer symbolischen Struktur werden. Die leichte Konstruktion besticht durch ihre Einfachheit und klar formulierte Vision. Sie besteht aus einem von Stahlpfeilern getragenen Rundgang. Dieser ist mit einer Brüstung versehen, die hier und da Sichtöffnungen aufweist. Der Stil der Struktur ist beispielhaft für Bauwerke in anfälligen Lebensräumen und ergründet ganz bewusst drei Themen: Kontrast ohne Gegensatz, Zugänglichkeit ohne Kontakt und Monumentalität ohne Gewicht.

Cette structure symbolique des environs de Sydney est un espace d'exposition qui associe divers éléments antagonistes : protection de l'environnement, accès au public et standards de sécurité suffisants. L'ouvrage, parfaitement transparent, se caractérise à la fois par une beauté toute simple et la poésie visionnaire qui en émane. Il se compose d'une plateforme circulaire qui repose sur des piliers en acier et est dotée d'une rambarde ajourée offrant des vues plongeantes sur les environs. Parfaitement adaptée aux biotopes fragiles, la structure a été conçue en suivant trois grandes lignes : contraste sans contradiction, accessibilité sans contact et monumentalité sans poids écrasant.

From left to right, from above to below:
Railing, shaded interactive section, elevation.
Right: Meeting point.

Von links nach rechts, von oben nach unten:
Geländer, verschatteter interaktiver Bereich, Ansicht.
Rechts: Treffpunkt.

De gauche à droite, du haut vers le bas:
Main courante, espace d'exposition interactif
ombragé, vue d'ensemble avec reflet dans l'eau.
À droite: Point de rencontre.

YBBS BRIDGE,
AMSTETTEN, AUSTRIA

GERHARD KOPEINIG ARCH + MORE
WITH RALF STERNIG

www.arch-more.com, www.zt-sternig.at

Type: Suspension bridge, **Length:** 92 m, **Completion:** 2007, **Photos:** Walter Luttenberger / Gratkorn.

Left: Perspective. Links: Perspektive. À gauche: Vue en perspective. | Right: Section. Rechts: Querschnitt. À droite: Vue en coupe.

The new Ybbs bridge is located in a natural setting that is only disrupted by an existing railway bridge. The newly built bridge features two prominent 13-meter high frames, which serve pedestrians and bicyclists as the "gate to the bridge." These steel gates carry the delicate roadway that is suspended from cables where visitors can feel the energy of the river and experience the tension between the two banks. Made of massive welded steel sheets and resting on ferroconcrete foundations the size of houses, the gate frames were coated in anthracite grey. A seven-centimeter steel cable carries the roadway, which is framed by a stainless steel mesh railing on both sides.

Die neue Ybbsbrücke befindet sich inmitten von Natur, deren Anblick lediglich durch eine bestehende Eisenbahnbrücke gestört wird. Am Neubau dienen zwei markante, 13 Meter hohe Uferrahmen Fußgängern und Radfahrern als „Tor zur Brücke". Diese Tore aus Stahl tragen die filigrane, an Seilen hängende Fahrbahn. Besucher können hier die Energie des Flusses spüren und die Spannung zwischen den Ufern erleben. Die Torrahmen aus massiven Stahlblechen geschweißt und auf Stahlbetonfundamenten in der Größe von Wohnhäusern ruhend, wurden Anthrazitgrau beschichtet. Ein sieben Zentimeter dickes Stahlseil trägt die seitlich durch ein Edelstahlnetzgeländer begrenzte Fahrbahn.

Cette nouvelle passerelle sur l'Ybbs se trouve en pleine nature, le seul élément construit à proximité étant un pont de chemin de fer. Le tablier est suspendu à deux pylônes gris anthracite de treize mètres de haut. Réalisés en plaques d'acier soudées et fixés à des culées en béton armé de la taille d'une maison, les pylônes constituent en quelque sorte les « portes du pont ». Lorsqu'ils traversent la passerelle, piétons et cyclistes peuvent apprécier l'énergie de la rivière qui coule au-dessous d'eux. Des câbles en acier de sept centimètres de diamètre supportent le poids du tablier et de son parapet en grillage d'acier inox.

Perspective by night. Perspektive bei Nacht. Passerelle la nuit.

From left to right, from above to below:
The distinctiv gate frames, elevation, view to the substructure and ferroconcrete fundament, gate frame.
Right: View of the woods through the gate frame.

Von links nach rechts, von oben nach unten:
Das markante „Tor zur Brücke", Ansicht, Blick auf die Unterkonstruktion und die Stahlbetonfundamente, Brückentor.
Rechts: Blick durch das Tor Richtung Wald.

De gauche à droite, du haut vers le bas:
Le « pylône-porte », vue du tablier, tablier vu d'en dessous et culée en béton armé, autre vue d'un pylône.
À droite: Vue dans l'axe de la passerelle avec la forêt à l'arrière plan.

SCHANERLOCH BRIDGE,

EBNIT / DORNBIRN, AUSTRIA

MARTE.MARTE ARCHITEKTEN

www.marte-marte.com

Type: Arched bridge, **Length:** 23 m, **Completion:** 2005, **Photos:** Marc Lins.

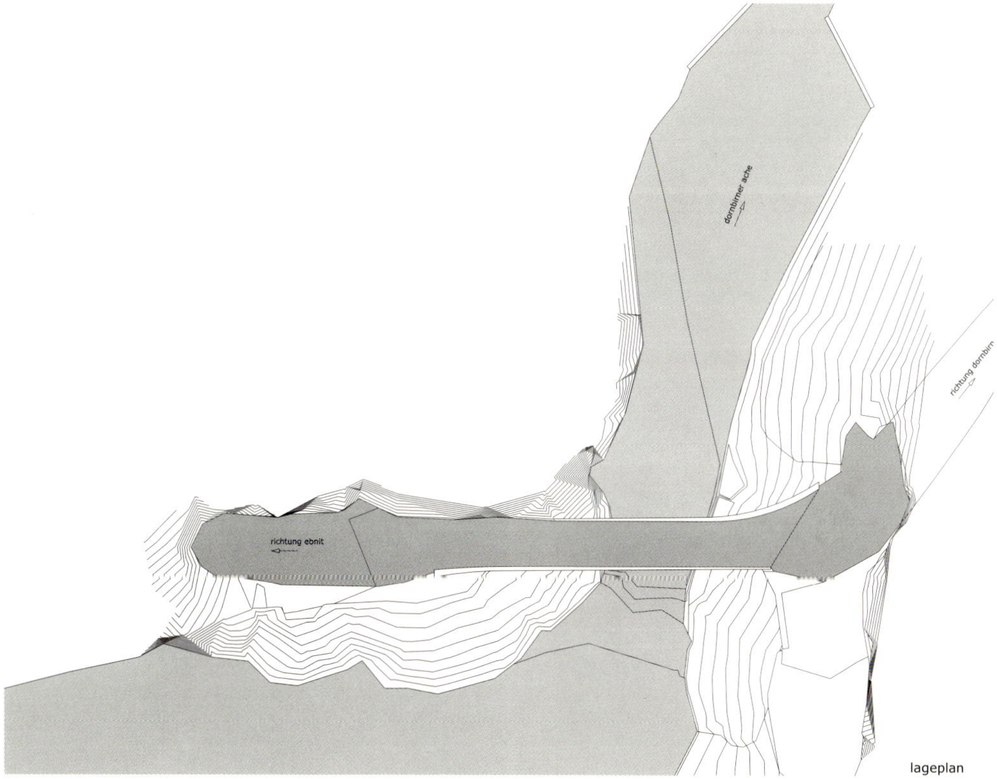

Left: Elevation. Links: Ansicht. À gauche: Vue d'ensemble. | Right: Site plan. Rechts: Lageplan. À droite: Plan de situation.

The bridge through the Schanerloch gorge is part of the mountain road from Dornbirn to the hamlet of Ebnit situated by the source of the river Dornbirner Ache at the foot of scenic mountains. Based on the typology of a stone arch bridge, modern technology takes the geometry of the arch to its very limits. The reduction of the arch rise to a statically necessary minimum is combined with a twist along one axis. The latter also responds to the shape of the road that curves immediately following the bridge in both directions. The result is a concrete sculpture that unveils its compelling fascination most when seen from the shore of the river winding through the gorge.

Die Brücke durch die Schanerloch Schlucht ist Teil der Bergstraße von Dornbirn zum Dörfchen Ebnit, einer ehemaligen Walsersiedlung am Ursprung der Dornbirner Ache am Fuße einer malerischen Berglandschaft. Basierend auf der Typologie der Steinbogenbrücke ist sie das Ergebnis einer Ausreizung der Bogenform mit modernen Technologien. Der Bogen wurde auf ein statisch notwendiges Minimum reduziert und zusätzlich einer Verwindung unterzogen. Diese reagiert auf den beidseitig in Krümmung wegziehenden Straßenverlauf. Das Resultat ist eine Betonskulptur, die ihre fesselnde Faszination am besten preisgibt, wenn man sie vom Ufer des Flusses betrachtet, der sich durch die Schlucht windet.

Un nouveau pont permet de relier les villages de Dornbirn et Ebnit, localité que les Walsers ont fondée au Moyen Âge dans un site grandiose où le torrent Dornbirner Ache prend sa source. Il s'agit d'un ouvrage moderne inspiré des ponts voûtés traditionnels en pierres. La voûte, réduite ici à un minimum, se complète par une torsion axiale adaptée aux virages que fait la route sur les deux rives. Le pont est ainsi une véritable sculpture en béton, qu'on apprécie tout particulièrement au niveau du torrent qui se fraye un passage dans les gorges.

From left to right, from above to below:
View from below, perspective,
curve of the bridge, elevation.
Right: On the bridge.

Von links nach rechts, von oben nach unten:
Blick auf die Unterseite, Perspektive,
Schwung der Brücke, Ansicht.
Rechts: Auf der Brücke.

De gauche à droite, du haut vers le bas:
Le pont vu d'en dessous, deux vues
latérales, le pont vu du torrent.
À droite: Vue du tablier.

DANUBE BRIDGE,
LINZ, AUSTRIA

GMP - ARCHITEKTEN VON GERKAN, MARG UND PARTNER / MAREK NOWAK, MICHAEL KÖNIG

www.gmp-architekten.de

Type: Suspension bridge, **Length:** 300 m, **Design:** Volkwin Marg and Jörg Schlaich, **Structural engineers:** Schlaich Bergermann und Partner, **Completion:** 2015, **Photos:** Courtesy of gmp Architekten (32, 34 a. r.), Gärtner + Christ, Hamburg (34 a. r., 34 b.), Bünck+Fehse, Berlin (35).

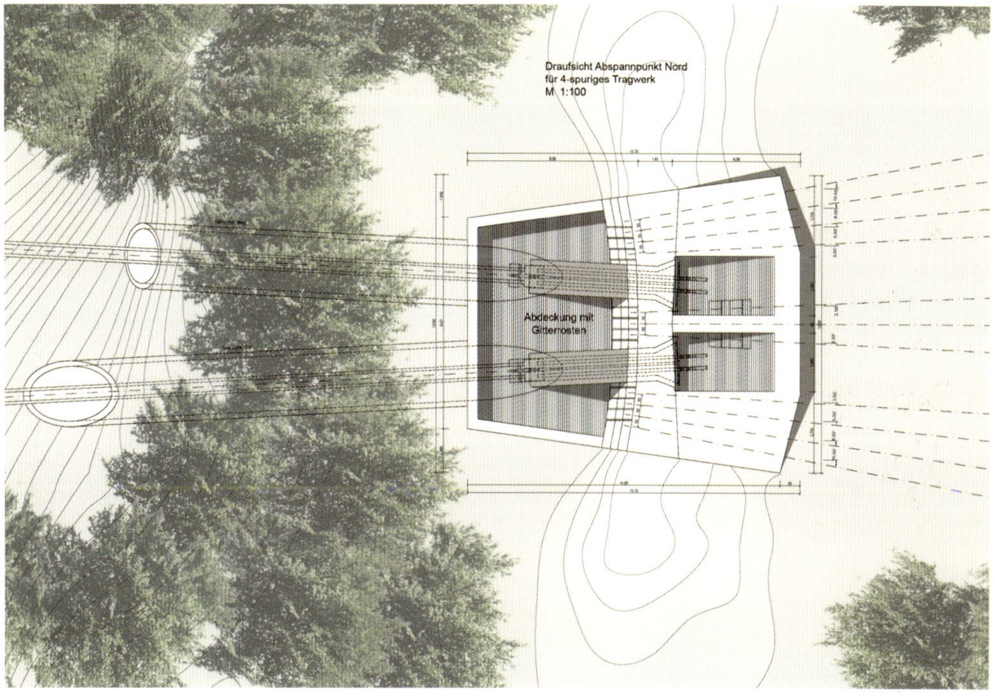

Left: View along the Danube. Links: Blick entlang der Donau. À gauche: Vue du pont sur le Danube. | Right: Northern tensioning base. Rechts: Nördliches Abspannfundament. À droite: Ancrage nord.

A key criterion for this design was maintaining an unobstructed view of the beautiful Danube landscape surrounding the bridge. A suspension bridge was found to be the ideal solution for the topography, which is dominated by bluffs, with suspension cables fixed to the minimally fissured rocks. Two separate parallel motorways, which continue in tunnels inside the bluffs, are suspended by vertical cables with two bearing cables each. Two-meter high steel box girders serve as stiffening girders in conjunction with a concrete slab. This renders the centrally arranged steel boxes invisible.

Wichtig für den Entwurf dieser transparenten Brücke war die schöne Donaulandschaft, die möglichst unbeeinträchtigt wahrgenommen werden sollte. Ideal für die von Steilufern bestimmte Topographie ist eine Hängebrücke, deren Hängeseile im kaum zerklüfteten Felsgestein halt finden. Zwei getrennte, parallele Fahrbahnen, die in den Steilufern in Tunneln weitergeführt werden, sind mit vertikalen Hängeseilen an jeweils zwei Tragseilen abgehängt. Die beiden Tragseile jeder Brückenhälfte werden zu einem Verankerungspunkt zusammengeführt. Als Versteifungsträger dienen zwei Meter hohe Stahlhohlkästen im Verbund mit einer Betonplatte. Die mittig angeordneten Stahlkästen verschwinden dabei optisch.

Les architectes ont ici choisi une forme filigrane afin de minimiser l'impact sur l'environnement. Un pont suspendu s'imposait par ailleurs du fait de la topographie et des berges abruptes du Danube. Les deux chaussées parallèles et indépendantes du tablier, qui se prolongent par des tunnels de chaque côté du fleuve, sont soutenues par des haubans verticaux accrochés à quatre grands câbles incurvés. La rigidité de l'ensemble est assurée par une grande dalle en béton et des caissons en acier de deux mètres de haut, invisibles du fait de leur position centrale.

From left to right, from above to below:
Tunnel entrance, night view of bridge rendering,
view of the bridge with tensioning from above.
Right: Front elevation of the bridge over the Danube.

Von links nach rechts, von oben nach unten:
Tunneleingang, Brückensimulation bei Nacht,
Brückenaufsicht mit Verankerung.
Rechts: Frontalansicht der Brücke über die Donau.

De gauche à droite, du haut vers le bas:
Extrémité du pont et tunnels, le pont la nuit,
vue à la verticale du pont.
À droite: Vue du pont sur le Danube.

SEEROSENBRÜCKE,
TULLN AN DER DONAU, AUSTRIA

ARCHITEKTURSTUDIO
BULANT & WAILZER

www.austria-architects.com/bulant_wailzer

Type: Girder bridge, **Length:** 140 m, **Architects:** Aneta Bulant-Kamenova and Klaus Wailzer, **Structural engineers:** Vasko + Partners Ingenieure, Vienna / Lothar Heintrich, **Completion:** 2008, **Photos:** Rupert Steiner, Vienna, Archiv Bulant & Wailzer/ Foto Fa. Carl Stahl (38 b. l.).

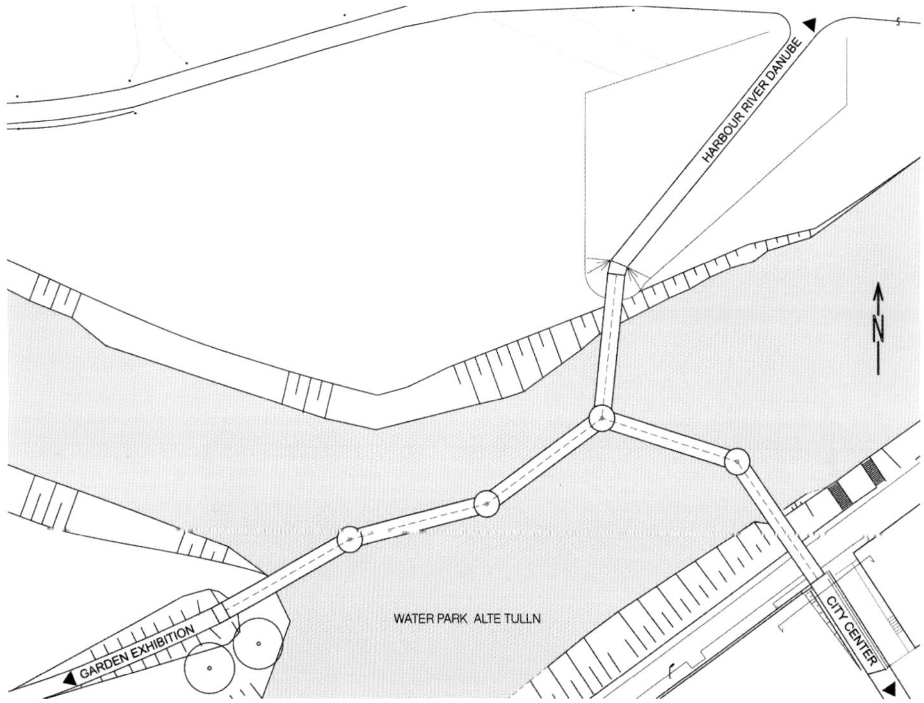

Left: Route system with branchings and changes of direction. Links: Wegeführung mit Verzweigung und Wechseln der Richtung. À gauche: Passerelle allant dans plusieurs directions. | Right: Site plan. Rechts: Lageplan. À droite: Plan de situation.

"Metabolically" branched, the Seerosen (water lily) bridge extends across the water like a delicate spider's web. Inspired by the shape of water lilies, the piles are made of ready-made concrete elements and connected to each other by consistently similar steel constructions. The direction of travel can be changed at will at any hub point. The height difference between the end points is almost three meters. However, instead of being inclined in one direction across the entire length, the bridge offers a pleasant wave-like movement with variously inclined ramps. The greater height required at the center of each single-span girder is applied towards the top and in line with the desired ramp design.

„Metabolistisch" verzweigt, überspannt die Seerosenbrücke die Wasserfläche wie ein leichtes Spinnennetz. Inspiriert von der Seerosenform sind die Stützen aus Betonfertigteilen ausgeführt und mit immer gleichen geraden Stahlkonstruktionsteilen verbunden. Die Laufrichtung kann in jedem Drehpunkt beliebig gewechselt werden. Der Höhenunterschied zwischen den Endpunkten beträgt fast drei Meter. Statt nur in einer Richtung über die Gesamtlänge geneigt zu sein, bietet die Brücke eine angenehme wellenförmige Bewegung mit verschieden geneigten Rampen. Die notwendige größere Höhe in der Mitte der jeweiligen Einfeldträger ist nach oben aufgetragen und günstig zur gewünschten Rampenausbildung verwendet.

Cette passerelle présente la particularité « métabolique » de bifurquer au-dessus de l'eau. Elle se compose de piles en béton préfabriquées inspirées de la forme des nénuphars, ainsi que d'éléments en acier rectilignes qui sont tous de construction similaire et partent systématiquement dans des directions différentes à l'endroit où ils reposent sur les piles. La différence de hauteur entre les têtes de ponts va jusqu'à trois mètres. Au lieu de la compenser par une rampe linéaire, le tablier monte et descend de manière à former une ondulation harmonieuse. De plus, les surfaces inférieures et supérieures des éléments du tablier ne sont pas parallèles afin d'en renforcer la rigidité, ce qui augmente encore l'impression d'ondulation.

From left to right, from above to below:
Aerial view, spots for resting and viewing,
angular deck with varying heights, on the bridge.
Right: Piles and joining members from below.

Von links nach rechts, von oben nach unten:
Luftaufnahme, Ruhe- und Aussichtspunkte,
winkelige Deckform mit variierender Höhe, auf der Brücke.
Rechts: Stützen und Verbindungselemente von unten.

De gauche à droite, du haut vers le bas:
Passerelle vue de haut, espaces circulaires
intermédiaires, trajet sinueux, vue du tablier.
À droite: Piles et éléments du tablier vus d'en dessous.

LUGNER BRIDGE,

VIENNA, AUSTRIA

ARCHITEKTURSTUDIO BULANT & WAILZER

www.austria-architects.com/bulant_wailzer

Type: Cable-stayed bridge, **Length:** 40 m, **Architects:** Aneta Bulant-Kamenova and Klaus Wailzer, **Structural engineers:** Vasko + Partners Ingenieure (Lothar Heinrich), **Light design:** Bartenbach Lichtlabor, **Steel and glass construction:** Waagner – Biro, **Completion:** 2005, **Photos:** Rupert Steiner, Vienna.

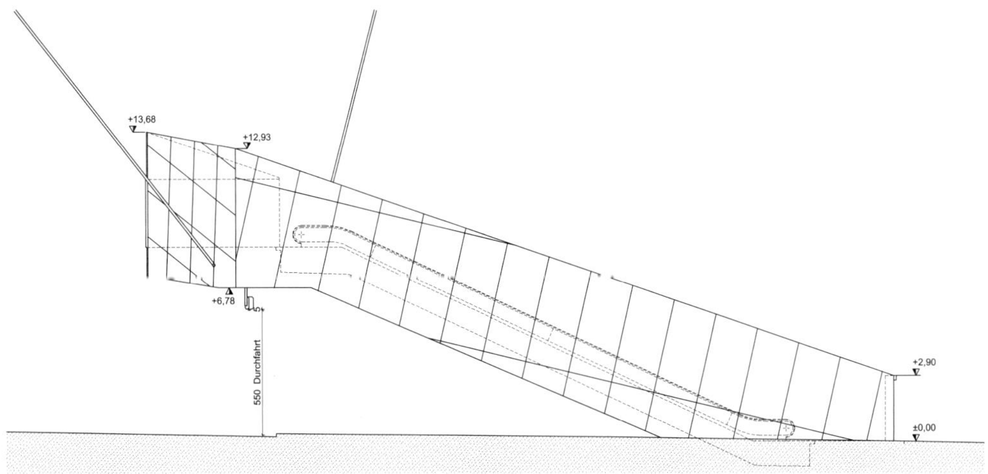

Left: View from north. Links: Blick von Norden. À gauche: Passerelle vue du nord. | Right: Section. Rechts: Schnitt. À droite: Vue en coupe.

The bridge is suspended on two lateral cables from the building of the "Lugner City" shopping center across the "Gürtel-Boulevard". An escalator takes pedestrians up to the platform with a panorama café. The distinguishing feature of the bridge is the ingenious relationship between the frame, structural geometry, and the cover. The glass cover is folded across the body like Origami. The lateral sides of the bridge are not parallel but slightly offset to each other, creating a unique spatial impression with an altered perspective. The counterpoint superimposition of the bridge body and glass cover is another attractive feature.

Die Brücke hängt an zwei Schrägseilen vom Gebäude des Einkaufzentrums „Lugner City" über den „Gürtel-Boulevard". Eine Rolltreppe führt Fußgänger hinauf auf die Plattform mit Panoramacafé. Das Besondere an der Brücke ist die raffinierte Beziehung zwischen Tragwerk, Baukörpergeometrie und Hülle. Die Glashülle ist wie Origami über den Baukörper gefaltet. Die Seitenflächen der Brücke sind nicht parallel, sondern leicht gegeneinander verschwenkt, wodurch sich im Inneren ein eigenwilliger, perspektivisch veränderter Raumeindruck ergibt. Die kontrapunktische Überlagerung der Systeme von Baukörper und Glashülle hat Ihren besonderen Reiz.

Cette passerelle est accrochée par deux câbles obliques au bâtiment principal du centre commercial « Lugner City ». Un escalator assure la liaison entre la rue et une plate-forme avec café panoramique. L'ouvrage d'art se caractérise par des rapports subtils entre la structure porteuse d'une part, la géométrie et la consistance de l'enveloppe de type origami d'autre part. Les parois ne sont pas parallèles, ce qui génère à l'intérieur de la passerelle une perspective particulière. Le contraste entre la structure et l'enveloppe est particulièrement esthétique.

From left to right, from above to below:
Escalators, panorama café on bridge platform, reflections on escalator façade during night, view from south.
Right: View from south-east.

Von links nach rechts, von oben nach unten:
Rolltreppe, Panoramacafé auf der Brückenplattform, nächtliche Spiegelungen auf der Rolltreppenfassade, Südansicht.
Rechts: Blick von Südosten.

De gauche à droite, du haut vers le bas:
Escalator, café panoramique, deux vues de la passerelle la nuit.
À droite: Passerelle vue du sud-est.

SKYWALK SPITTELAU,
VIENNA, AUSTRIA

ARGE BULANT-WAILZER / WAGNER / FRITSCH

www.austria-architects.com/bulant_wailzer

Type: Continuous girder bridge, **Length:** 120 m, **Architects:** Aneta Bulant-Kamenova and Klaus Wailzer. **Structural engineers:** Karlheinz Wagner, Rudolf Fritsch, **Completion:** 2007, **Photos:** Rupert Steiner, Vienna, Archiv Bulant & Wailzer/MirAlf, Vienna (46 b. l.).

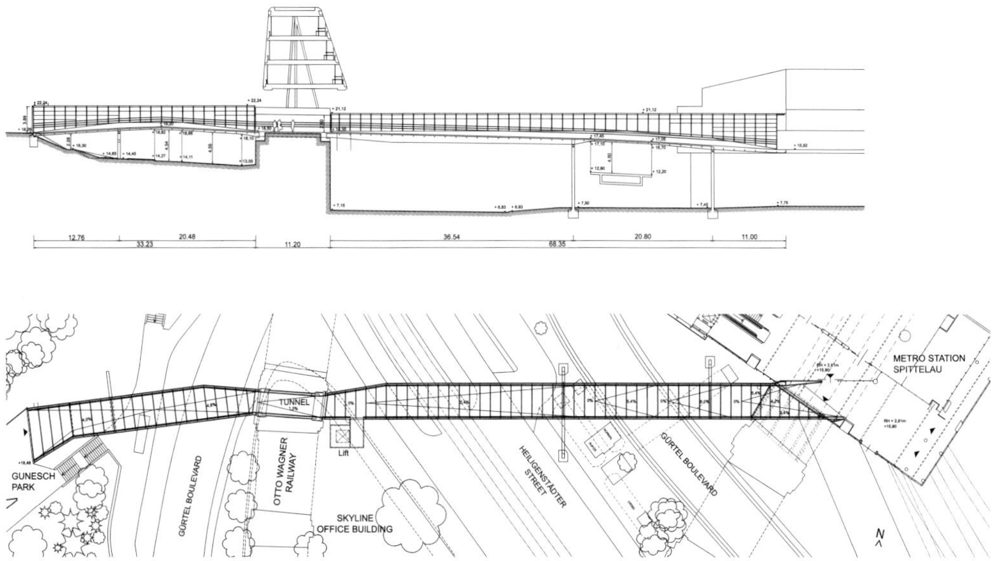

Left: Over the Heiligenstädterstrasse by night. Links: Nachts über der Heiligenstädterstraße. À gauche: Passerelle la nuit. | Right: Floor plan and section. Rechts: Grundriss und Schnitt. À droite: Vue en coupe et plan.

The architecture of the Skywalk was exclusively developed from the complex urban environment in a simplified and reduced form. The components are "parametrically" defined by each other. The height difference of the crossing streets define the ramps of the bridge floor, the power flow defines the height changes of the pillars, the width of the tunnel through the Otto-Wagner urban rail arches determines the changes of the bridge's width and direction. A key architectural feature is the horizontal glass roof of the housing, through which the architectural sculpture is integrated into the heterogeneous surroundings. This results in differentiated areas on the inside with various ceiling heights.

Die Architektur des Skywalk wurde ausschließlich aus dem komplizierten urbanen Umfeld in vereinfachter und reduzierter Form entwickelt. Die Komponenten sind „parametrisch" voneinander bestimmt. Die Niveauunterscheide der querenden Straßen bestimmen die Rampen des Brückenbodens, der Kraftfluss bestimmt die Höhenveränderung der Träger, die Breite des Tunnels durch die Otto-Wagner-Stadtbahnbögen verursacht den Wechsel der Brückenbreite und -richtung. Eine wesentliche architektonische Entscheidung ist das horizontale Glasdach der Einhausung, wodurch sich die Bauplastik einheitlich in die heterogene Umgebung eingliedert. Im Innenraum entstehen so differenzierte Bereiche mit verschiedenen Raumhöhen.

Le style simpliste et réducteur de cette passerelle a été choisi de manière délibérée afin de former un contraste dans un environnement urbain compliqué. Les composantes de l'ouvrage sont des paramètres en interaction : les différences de niveau des rues en contrebas expliquent l'inclinaison du tablier, les forces en action déterminent la différence de hauteur des supports, le passage sous la ligne de chemin de fer urbain est l'occasion de modifier la trajectoire générale de l'ouvrage. L'enveloppe en verre unifie l'ensemble de la passerelle dans un contexte architectural particulièrement hétérogène. L'intérieur présente quant à lui divers espaces plus ou moins hauts de plafond.

From left to right, from above to below:
Skywalk to Skyline office building, interior,
crossing the street, view from the street.
Right: Panorama by night.

Von links nach rechts, von oben nach unten:
Skywalk zum Bürogebäude Skyline, Innenaufnahme,
Überquerung der Straße, Ansicht von der Straße.
Rechts: Nachtpanorama.

De gauche à droite, du haut vers le bas:
Passerelle « Skywalk » et immeuble de
bureaux « Skyline », intérieur, passerelle
vue de haut, passerelle vue d'en bas.
À droite: Panorama nocturne.

MARIENSTEG,

WERNSTEIN / AUSTRIA, NEUBURG / GERMANY

ERHARD KARGEL

www.kargel.co.at

Type: Suspension bridge, **Length:** 150 m, **Light artist:** Waltraut Cooper, **Client:** City of Wernstein am Inn, City of Neuburg am Inn, **Completion:** 2006, **Photos:** Erhard Kargel, Linz.

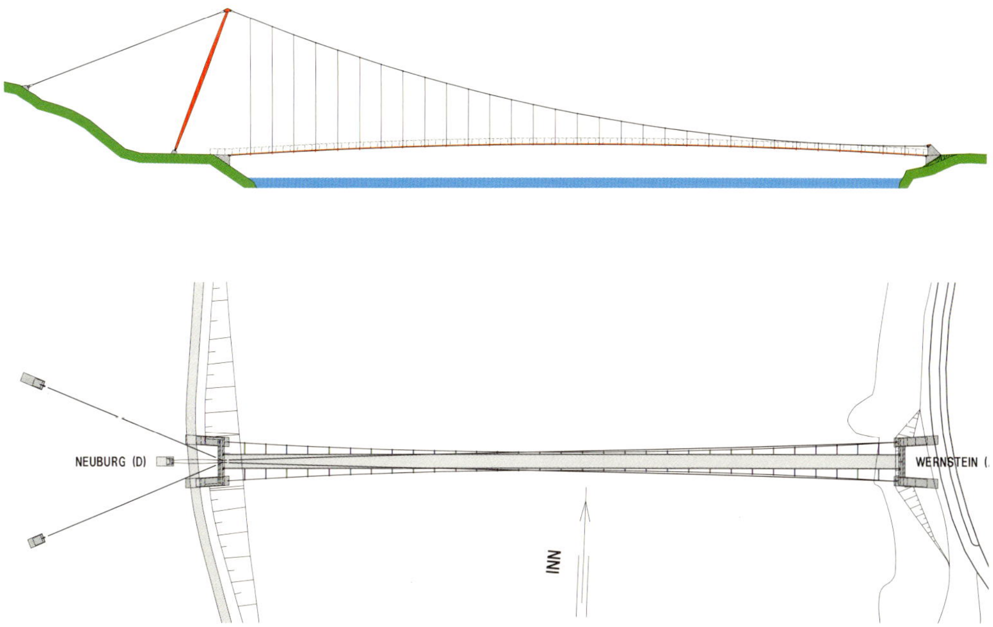

Left: On the footbridge to Neuburg. Links: Auf dem Steg nach Neuburg. À gauche: La passerelle en direction de Neuburg. | Right: Sections and floor plan. Rechts: Ansicht und Grundriss. À droite: Élévation et plan.

A connection across the Inn, which has developed in the course of several centuries, is now firmly established through the Mariensteg. The asymmetrical shape of the 144-meter wide suspended bridge with a single pylon emerged from the surrounding landscape and location. To increase the rigidity of the extremely slim construction, two wind cables were introduced next to the deck. Their spatial curvature and prestressing allow them to absorb horizontal and vertical forces. This concept allows for extremely simple assembling and potentially disassembling of the bridge. The bridge features a minimalist design with low resource requirements. It is difficult to conceive of a thinner construction.

Eine über Jahrhunderte gewachsene Verbindung über den Inn wird durch den Mariensteg gefestigt. Die asymmetrische Form der 144 Meter weit gespannten Hängebrücke mit nur einem Pylon ist eine Reaktion auf das Landschafts- und Ortsbild. Zur Erhöhung der Steifigkeit der äußerst schlanken Konstruktion wurden neben der Fahrbahn zwei Windseile eingeführt. Ihre räumliche Krümmung und Vorspannung erlaubt ihnen, horizontale und vertikale Kräfte aufzunehmen. Dieses Konzept ermöglicht eine denkbar einfache Montage und gegebenenfalls auch Demontage. Die Brücke ist ein minimalistisch gestaltetes Bauwerk mit geringem Ressourcenverbrauch. Eine weitere Ausdünnung ist kaum vorstellbar.

Une passerelle récemment construite simplifie considérablement les rapports transfrontaliers qui existent depuis des siècles entre Wernstein (Autriche) et Neuburg (Bavière). Afin de minimiser l'impact sur le paysage, cet ouvrage asymétrique et particulièrement svelte de 144 mètres de long est suspendu à un unique pylône par l'intermédiaire de deux câbles qui bordent le tablier et compensent les poussées horizontales et verticales. Cette conception a considérablement simplifié la réalisation de l'ouvrage, tout en rendant possible un éventuel démontage. Il s'agit là d'une passerelle minimaliste, tant par son aspect que par la quantité de ressources utilisées. Il eût été difficile de faire moins.

From left to right, from above to below:
Footbridge with Neuburg, transparency resulting from perforated grillage, view with Neuburg, needle-shaped pylon.
Right: Excellent incorporation into the landscape.

Von links nach rechts, von oben nach unten:
Steg mit Neuburg, Transparenz durch perforierten Rost, Ansicht mit Neuburg, nadelförmiger Pylon.
Rechts: Hervorragende Einbindung in die Landschaft.

De gauche à droite, du haut vers le bas:
Vue de la passerelle et du château de Neuburg, tablier transparent en grillage, autre vue de la passerelle et du château de Neuburg, vue du pylône filiforme.
À droite: Intégration parfaite au paysage.

XIAN INTERNATIONAL HORTICULTURAL EXPO MASTER PLAN,

XIAN, CHINA

www.serie.co.uk

Type: Arched bridge, **Length:** 1,000 m, **Completion:** Ongoing, **Renderings:** Courtesy of Serie Architects.

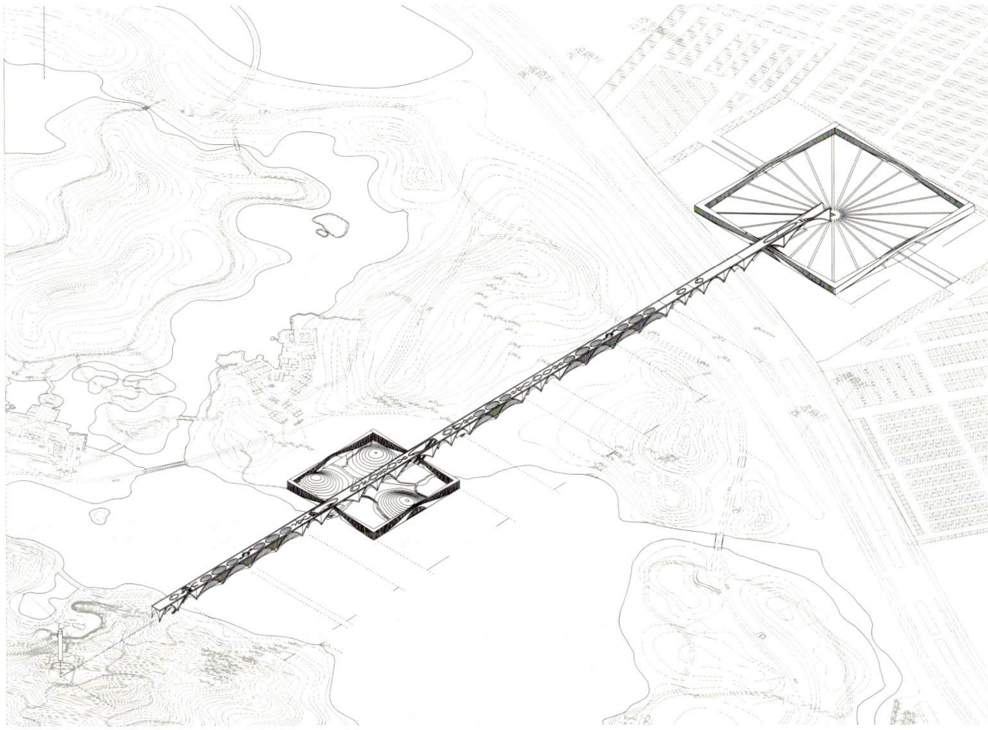

Left: Green houses, seen from park and lake. Links: Blick von Park und See auf die Gewächshäuser. À gauche: Les serres vues du lac. | Right: Master plan. Rechts: Masterplan. À droite: Schéma directeur.

Five Climates Crossing is a strong and clear structure serving as a main organizing element for the master plan of the Xian Expo Park. This one-kilometer linear structure is made up of five green houses. Like the Xian city wall, it will mark the center of the park and simultaneously act as a connecting element, linking the entrance square to the north, Chang'an Park in the middle, and the viewing tower to the south. The green houses are arranged linearly as five different climatic zones, allowing visitors to move sequentially from one green house to another while enjoying the view of the park.

Five Climates Crossing ist ein klar formuliertes Bauwerk mit starkem architektonischen Ausdruck, das dem Gesamtplan des Xian Expo Parks als strukturierendes Element dient. Die einen Kilometer lange lineare Struktur besteht aus fünf Gewächshäusern. Ähnlich der Stadtmauer Xians wird das Bauwerk das Zentrum des Parks bestimmen und gleichzeitig eine Verbindung zwischen dem Eingangsbereich im Norden, dem Chang'an Park in der Mitte und dem Aussichtsturm im Süden herstellen. Im Inneren sind die Gewächshäuser der Länge nach in fünf Klimabereiche unterteilt. Besucher können sich von einem Gewächshaus zum anderen bewegen und gleichzeitig die Ausblicke auf den Park genießen.

La construction de cet ouvrage appelé « Five Climate Crossing » est prévue par le schéma directeur pour l'exposition de Xian et son parc. Il s'agit d'un bâtiment linéaire d'un kilomètre de long orienté nord/sud, qui va de l'entrée du parc à une tour panoramique en passant par le Jardin Chang'an, et qui abrite cinq serres reproduisant les conditions climatiques de cinq régions différentes. Simple et clair, cet élément fortement structurant est en harmonie avec la tradition architecturale et urbanistique locale, notamment les anciennes fortifications de la ville.

From left to right, from above to below:
Crossing, entrance square on the north,
bridge with Chang'an Park at the center.
Right: Greenhouse interior.

Von links nach rechts, von oben nach unten:
Kreuzungspunkt, Eingangshof im Norden,
Brücke mit dem Chang'an Park im Zentrum.
Rechts: Im Gewächshaus.

De gauche à droite, du haut vers le bas:
« Crossing », entrée nord du parc,
vue aérienne avec au centre le Jardin Chang'an.
À droite: Intérieur de la serre.

SVRATKA RIVER FOOTBRIDGE,
BRNO, CZECH REPUBLIC

STUDIO ACHT WITH STRASKY, HUSTY AND PARTNERS

www.studioacht.cz, www.hsp.eu

Type: Arched stress ribbon bridge, **Length:** 42.90 m, **Structural engineers:** Strasky, Husty and Partners, **Completion:** 2007, **Photos:** Jiri Dvorak.

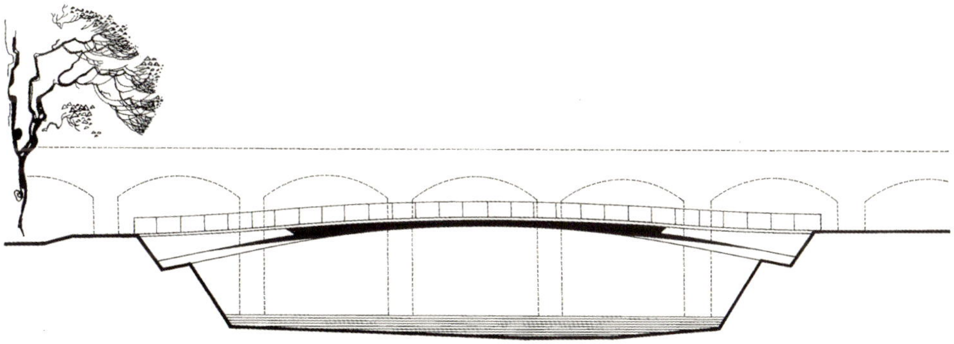

Left: Old and new arch bridge. Links: Alte und neue Bogenbrücke. À gauche: Passerelle et pont traditionnel à l'arrière plan. | Right: Section. Rechts: Schnitt. À droite: Vue en coupe.

The bridge connects a newly developed noble city district with the old city center. Near the new bridge an old multi-span arch bridge with piers in the river is situated. It was evident that the new bridge should also have an arch structure, but with a bold span without piers in the river bed. The deck of the bridge consists of a stress ribbon supported by a flat arch. Since the stress ribbon and the arch are fixed in common end abutments, the structure forms a self-anchored system that needs vertical supports only. The arch span is 42.9 meters with a rise of 2.65 meters. The stress ribbon and the arch are assembled of precast segments.

Die Brücke verbindet ein neu errichtetes prestigeträchtiges Viertel mit dem alten Stadtkern. Nahe der Brücke befindet sich eine alte Bogenbrücke mit mehreren Spannen und Pfeilern im Flussbett. Die neue Brücke sollte auch eine Bogenform haben, allerdings mit nur einer Spanne und ohne Pfeiler im Flussbett. Ein Zugband, welches ein flacher Bogen stützt bildet die Fahrbahn. Das Zugband und der Bogen sind in gemeinsamen Endfundamenten verankert wodurch die Struktur zu einem sich selbst tragenden System wird, das nur durch vertikale Träger gestützt wird. Die Bogenspanne beträgt 42,9 Meter, die Höhe 2,65 Meter. Das Zugband und der Bogen wurden aus vorgefertigten Teilen zusammengesetzt.

Cette passerelle relie le centre ville historique de Brno à un nouveau quartier résidentiel haut de gamme. La proximité d'un pont traditionnel à voûtes imposait de construire un ouvrage du même type, la technologie moderne permettant néanmoins d'élargir l'arche et de supprimer les piles intermédiaires. La passerelle se compose ainsi d'une arche supportant un tablier un peu moins cintré. Ces deux éléments préfabriqués sont ancrés dans des fondations communes en béton, de sorte que l'ensemble forme une structure autoporteuse. L'arche mesure 42,90 mètres de long, son point le plus haut se trouve à 2,65 mètres au-dessus du niveau de l'eau.

From left to right, from above to below:
Footbridge deck, overall view, railings.
Right: Stress ribbon and arch.

Von links nach rechts, von oben nach unten:
Auf der Fußgäbgerbrücke,
Gesamtansicht, Geländer.
Rechts: Spannband und Stützbogen.

De gauche à droite, du haut vers le bas:
Vue du tablier, vue d'ensemble,
détail de la rambarde.
À droite: Arche et tablier.

VLTAVA RIVER FOOTBRIDGE,
CESKE BUDEJOVICE, CZECH REPUBLIC

STRASKY, HUSTY AND PARTNERS

www.shp.eu

Type: Arched bridge, **Length:** 53.2 m, **Completion:** 2006, **Photos:** Jiri Dvorak.

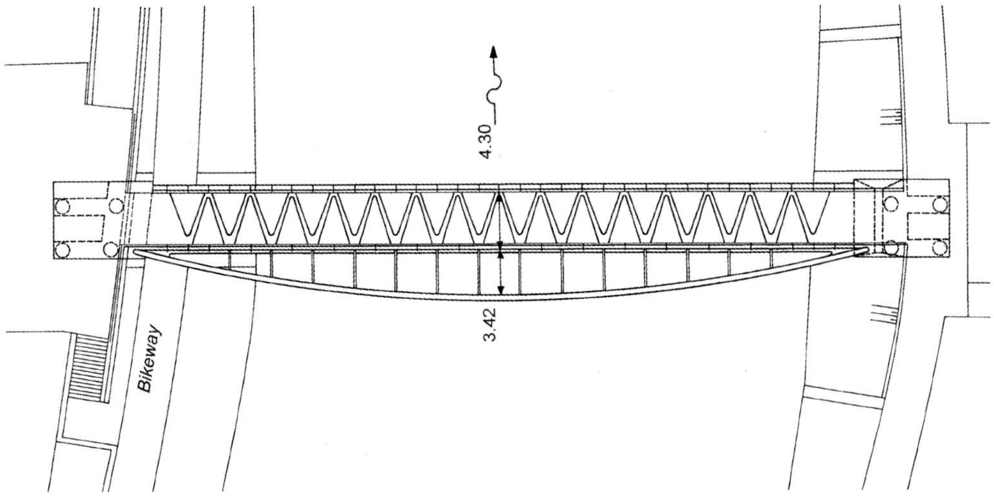

Left: View of the arch landing. Links: Brückenanlauf. À gauche: Tête de pont. | Right: Plan. Rechts: Plan. À droite: Plan.

Vltava River Footbridge is a part of newly built route network for pedestrians and cyclists, connecting the city's historical center with a new residential area built on the left bank. The structure is dominated by an inclined one-sided arch that is fixed into the deck. The span length is 53.2 meters; the maximum arch rise is eight meters. The structure is light and transparent, all structural components have a clean and simple shape. The arch is made of a steel pipe; the deck is formed by two edge pipes that are mutually connected by a horizontal truss and a composite concrete slab.

Die Vltava Fußgängerbrücke ist Teil eines neu errichteten Streckennetzes für Radfahrer und Fußgänger und verbindet das historische Stadtzentrum mit einem neuen Wohngebiet am linken Flußufer. Ihre Struktur wird von einem einseitigen, geneigten Bogen dominiert, der mit der Fahrbahn verbunden ist. Die Brücke überspannt 53,2 Meter und ihr Bogen hat eine maximale Höhe von acht Metern. Ihre Struktur erscheint leicht und transparent und alle strukturellen Komponenten haben eine klare und einfache Form. Der Bogen ist aus einem Stahlrohr gefertigt. Die Fahrbahn besteht aus zwei Rohren, die durch einen horizontalen Träger und eine Betonverbund-decke verbunden sind.

Cette passerelle pour piétons et cyclistes relie le centre ville historique de Ceske Budejovice à un quartier résidentiel nouvellement aménagé sur la rive gauche de la rivière Vltava. L'ouvrage se compose d'un tablier suspendu à une arche d'une portée de 53,20 mètres pour huit de haut. Des éléments aux formes claires et simples s'allient pour former une structure générale légère et transparente. L'arche consiste en un tube d'acier, le tablier en deux tubes servant de support à une dalle en béton.

From left to right, from above to below:
Handrails, arch and handrails,
general view, pillars.
Right: Illumination at night.

Von links nach rechts, von oben nach unten:
Geländer, Bogen und Geländer,
Gesamtansicht, Träger.
Rechts: Beleuchtung in der Nacht.

De gauche à droite, du haut vers le bas:
Rambarde, arche et rambarde,
vue d'ensemble, haubans.
À droite: Passerelle la nuit.

NEW KAROLINA FOOTBRIDGE,

OSTRAVA, CZECH REPUBLIC

EVA JIRICNA ARCHITECTS LTD.

www.ejal.com

Type: Arched suspension bridge, **Length:** 240 m, **Planning partners:** AED Project a.s Czech Republic, **Structural engineers:** Link Projekt s.r.o Czech Republic, **Client:** New Karolina Shopping Center Development s.r.o, **Completion:** 2011, **Photos:** EJAL.

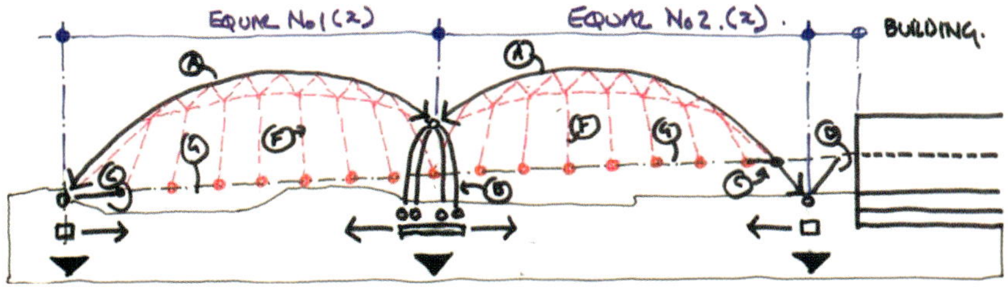

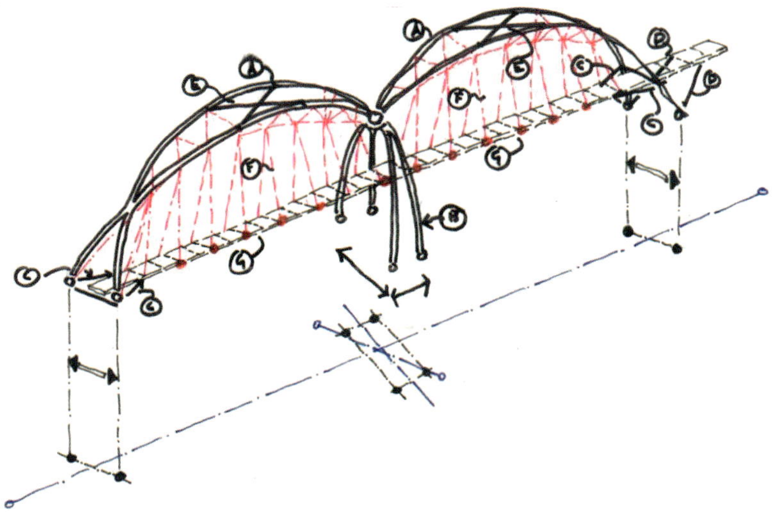

Left: Deck towards Karolina shopping center. Links: Brückendeck Richtung Einkaufszentrum Karolina. À gauche: Pont menant au centre commercial. | Right: Axonometric sketch. Rechts: Axonometrie-Skizze. À droite: Axonométrie.

Ostrava, an old mining town in the Czech Republic, is currently being transformed by a regeneration program replacing a declining mining industry. The first major development is a multifunctional shopping complex whose second floor is linked via the bridge's span of 240 meters to the existing city and its transport hubs. The arch design and intersection mirrors the pedestrian access points. The bridge is divided into two equal double-inclined suspension arches intersecting at the central transfer point – the symbolic meeting of old and new. This transfer arch provides lateral and vertical restraint to the six-meter wide suspended deck, constructed from a steel concrete composite.

Die ehemalige Bergbaustadt Ostrava in der Tschechischen Republik wird im Moment durch ein Erneuerungsprogramm von den Überresten der Bergbauindustrie bereinigt. Eines der ersten Bauvorhaben ist ein multifunktionaler Einkaufskomplex, dessen zweite Etage durch die Brücke mit ihren 240 Metern Spanne mit der Stadt und den Transportknotenpunkten verbunden wird. Die Bogenform und Schnittpunkte markieren die Fußgängerzugänge. Die Brücke ist in Zwillingsbögen unterteilt, die am Übergangspunkt, wo alt und neu verschmelzen, zusammentreffen. Dieser Übergangsbogen bildet die seitliche und vertikale Begrenzung für die sechs Meter breite Fahrbahn, die aus einem Stahl-Beton-Verbund gefertig ist.

Un plan de reconversion générale a été lancé à Ostrava, ancienne cité minière de République Tchèque. Il porte notamment sur la construction d'un centre commercial multifonctionnel dont le second étage doit être relié à la ville et à l'infrastructure des transports par un pont de 240 mètres de long. L'ouvrage se compose de deux arches dont l'intersection symbolise l'alliance de l'ancien et du nouveau, et d'un tablier suspendu de six mètres de large devant être réalisé en acier et béton.

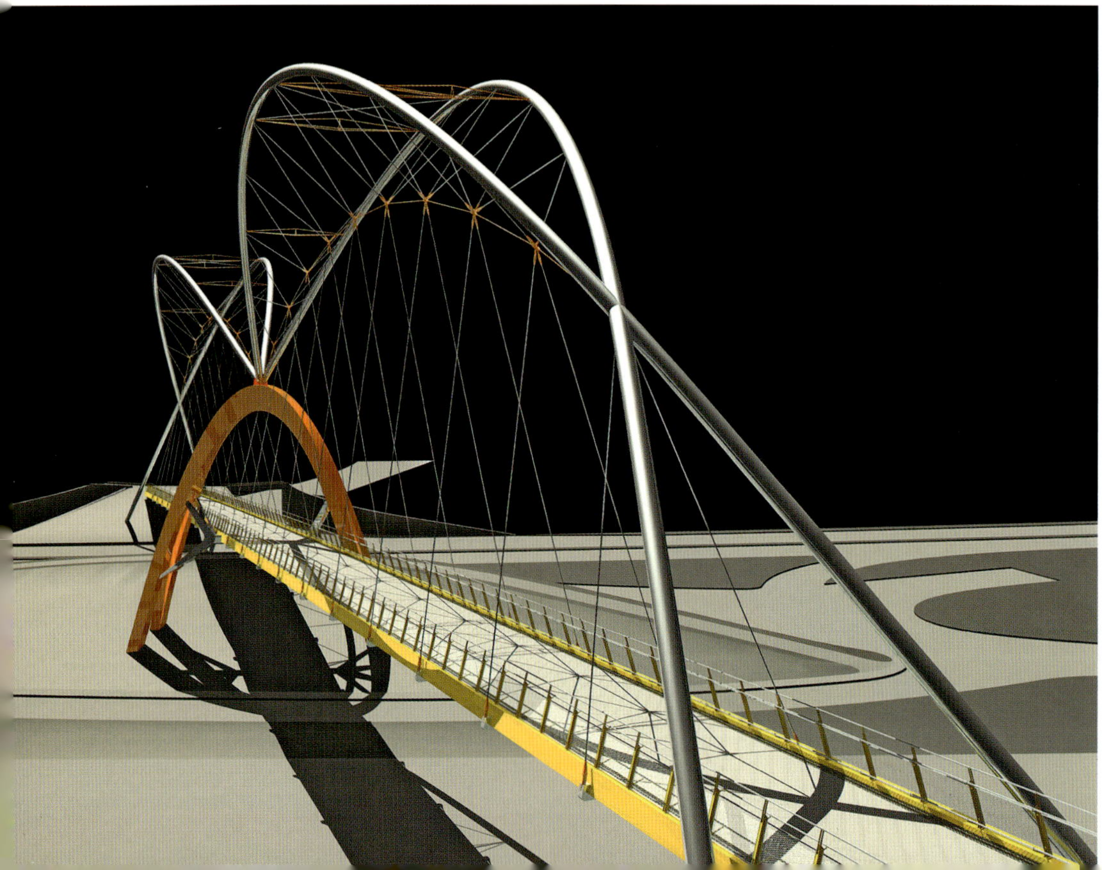

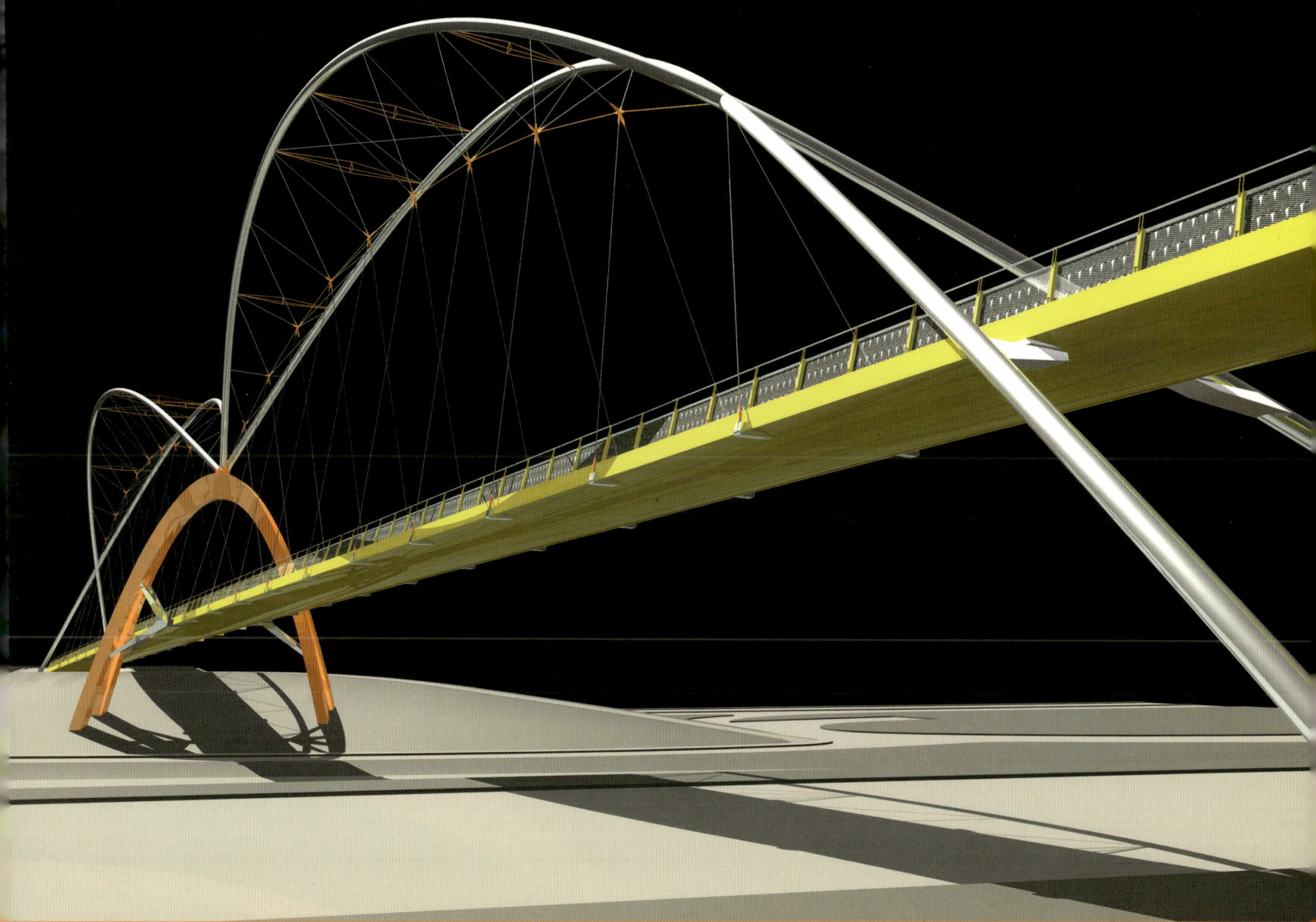

From left to right, from above to below:
View from Mistecka Street, view from Karolina
shopping center roof, view from shopping center.
Right: View from Porazkova Street.

Von links nach rechts, von oben nach unten:
Blick von der Mistecka Straße, Blick vom Dach des Karolina
Einkaufszentrums, Blick vom Einkaufszentrum.
Rechts: Blick von der Porazkova Straße.

De gauche à droite, du haut vers le bas:
Pont vu de la rue Mistecka, du toit du
bâtiment et du centre commercial.
À droite: Pont vu de la rue Porazkova.

BRIDGE ACROSS THE ODRA RIVER,
OSTRAVA, CZECH REPUBLIC

STRASKY, HUSTY AND PARTNERS

www.shp.eu

Type: Cable-stayed bridge, **Length:** 605 m, **Completion:** 2008, **Photos:** Stefan Spic (68, 70 a. r.), Jiri Dvorak (70 a. l., 70 b., 71).

Left: Main span across the Odra River. Links: Die Brücke überspannt den Fluss Odra. À gauche: Portée principale traversant l'Oder. | Right: Elevation. Rechts: Ansicht. À droite: Élévation.

Motorway D47 crosses the Odra River and Antosovice Lake between the cities of Ostrava and Bohumin at a skewed angle on a cable stayed bridge of 605 meters in length. Since the bridge is visible from different angles not only from the freeway, but also from the banks of the river and the lake, the designer wanted to create a clearly readable structure. As a result, the bridge deck is suspended in the bridge axis on a single 47-meter tall pylon. This way the stay cables with the deck present a clean structure in all directions. Both motorways are situated on separate structures that are mutually connected by the deck slab and the individual struts.

Die Schnellstraße D47 überquert die Oder und den Antosovicer See zwischen den Städten Ostrava und Bohumin im asymmetrischen Winkel auf einer Schrägseilbrücke mit 605 Metern Länge. Entwurfsziel war eine klar ablesbare Struktur, da die Brücke weithin sichtbar ist und das nicht nur von der Schnellstraße aus, sondern auch vom Flusslauf und vom Seeufer. Aus diesem Grund ruht die Fahrbahn in der Brückenachse auf einem einzelnen 47 Meter hohen Pylon. Auf diese Weise ergeben die Kabel und die Fahrbahnplatte eine klare Struktur in alle Richtungen. Beide Richtungsfahrbahnen ruhen auf separaten Konstruktionen, die durch die Fahrbahnplatte und durch die einzelnen Gestänge miteinander verbunden sind.

La route D47 franchit l'Oder et le lac d'Antosovice entre les villes d'Ostrava et de Bohumin au moyen d'un pont d'une longueur totale de 605 mètres. L'ouvrage, visible sous des angles différents à partir du fleuve et de ses berges, présente une structure claire et facilement lisible. La plus grande portée est suspendue par des haubans à un pylône de quarante-sept mètres de haut. Le reste du tablier à chaussées séparées repose sur des piles en béton.

From left to right, from above to below:
Pylon and stays, aerial view, approach spans.
Right: Approach spans viewed from below.

Von links nach rechts, von oben nach unten:
Pylone mit Verspannungen,
Luftaufnahme, Brücke über den Fluss.
Rechts: Brückenanlauf von unten.

De gauche à droite, du haut vers le bas:
Pylône et haubans, vue aérienne,
piles dans l'eau.
À droite: Vue des piles
supportant le tablier.

ØRESUNDSBROEN,

COPENHAGEN / DENMARK, MALMÖ / SWEDEN

ASO GROUP / ARUP WITH SETEC, GIMSING & MADSEN, ISC AND GEORG ROTNE

www.arup.com, www.gimsing.dk, www.isc.dk

Type: Cable-stayed bridge, **Length:** 7.85 km, **Completion:** 2000, **Photos:** Arup, London.

Left: Main span from the west. Links: Hauptspanne von Westen. À gauche: La grande portée vue de l'ouest. | Right: Drawing. Rechts: Zeichnung. À droite: Dessin d'artiste.

This huge bridge connecting Denmark and Sweden was commissioned by the two governments as a joint venture project. It is designed as a two level composite structure with the railway on the bottom - between two steel trusses - and the motorway on top. The bridge is the longest railway bridge in Europe and, with a span of 490 meters, the world's longest spanning cable-stayed bridge carrying both a road and a railway. With diagonals at 45 degrees, the truss girders have a more open bracing than usual. Within the cable-stayed bridge, the geometry is modified so that all diagonals have the same angle as the cables, which are anchored on the sides of the bridge by sloping brackets.

Diese gigantische Brücke verbindet Dänemark mit Schweden und wurde als Gemeinschaftsprojekt der beiden Regierungen in Auftrag gegeben. Als zweistöckige Struktur angelegt führt der Schienenstrang zwischen zwei Stahlträgern unterhalb der Fahrbahndecke entlang. Die Brücke ist Europas längste Eisenbahnbrücke und mit ihrer Spanne von 490 Metern zugleich die weltweit längste Schrägseilbrücke, die sowohl dem Schienen- als auch dem Individualverkehr dient. Die Fachwerkträger sind mit 45 Grad weiter gefächert als üblich. Innerhalb der Schrägseilbrücke ist die Geometrie so abgestimmt, dass alle Diagonalen die Winkel der Kabel aufgreifen, welche seitlich an geneigten Auslegern befestigt sind.

Ce pont gigantesque entre le Danemark et la Suède a été réalisé en commun par les deux pays. Il s'agit d'un ouvrage sur deux niveaux pour automobiles et trains. C'est à la fois le plus long des ponts de chemin de fer en Europe et celui au monde qui à la plus grande portée haubanée (490 mètres), capable de supporter le poids cumulé des voitures et d'un train. Ce qui explique l'angle inhabituel de 45 degrés formé par les haubans avec le tablier. Ceux-ci, en diagonale parfaite, sont fixés au tablier par des consoles inclinées.

From left to right, from above to below:
Bird's-eye view, distant view,
pylon structure, worm's-eye view of pylon.
Right: View along bridge.

Von links nach rechts, von oben nach unten:
Luftaufnahme, Ansicht aus der Ferne,
Pylone, Ansicht des Pylon von unten.
Rechts: Blick entlang der Brücke.

De gauche à droite, du haut vers le bas:
Vue aérienne, deux vues d'ensemble,
tablier vu d'en dessous.
À droite: Vue d'ensemble.

VIADUC DE MILLAU,
MILLAU, FRANCE

FOSTER + PARTNERS

www.fosterandpartners.com

Type: Cable-stayed bridge, **Length:** 2.46 km, **Client:** French Ministry of Equipment, Transport, Housing, Tourism and Sea, **Completion:** 2005, **Photos:** Eiffage CEVM / Foster + Partners / D. Jamme, **Plans:** Foster + Partners.

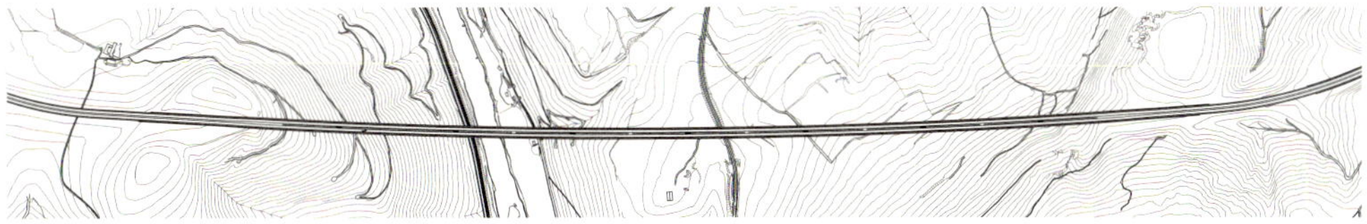

Left: Bridge across the valley of the Tarn. Links: Brücke über dem Tal des Tarn. À gauche: Le Tarn et Millau, avec le viaduc à l'arrière plan. | Right: View and layout. Rechts: Ansicht und Lageplan. À droite: Vue en élévation et plan de situation.

The bridge creates a direct route from Paris to Barcelona by crossing the River Tarn gorge. The cable-stayed, mast-supported structure is delicate and transparent with an optimum span between columns (342 meters). They range in height from 75 meters to 245 meters (larger than the Eiffel Tower), with the masts rising a further 87 meters above the road deck. Each column splits into two thinner more flexible columns below the roadway, forming an A-frame above deck level. The tapered form of the columns both expresses their structural loads and minimizes their profile. This gives the bridge a dramatic silhouette and, more crucially, minimizes intervention of the landscape.

Die Brücke quert die Schlucht des Flusses Tarn und schließt so eine Lücke in der Verbindung Paris-Barcelona. Ihre Ausführung als kabelgestützte, mastgetragene Konstruktion lässt sie elegant und transparent erscheinen. Die Stützen mit optimalem Abstand von 342 Metern sind von 75 bis 245 Meter hoch (höher als der Eiffelturm) und wurden oberhalb der Fahrbahn um weitere 87 Meter verlängert. Unterhalb der Fahrbahn teilen die Stützen sich auf und werden oberhalb von ihr in A-Form wieder zusammengeführt. Ihre sich verjüngende Form drückt die wirkenden Kräfte aus und minimiert zugleich ihre Präsenz. So entsteht eine dramatische Silhouette und, wichtiger noch, wird der Einfluss auf das Landschaftsbild minimiert.

Ce pont multi-haubané, qui franchit le Tarn, était le dernier tronçon manquant pour une liaison directe Paris-Barcelone par l'autoroute. La hauteur des piles varie : 77,5 mètres pour la plus petite, 245 mètres pour la plus haute. Les pylônes s'élèvent à 87 mètres au-dessus du tablier, de sorte qu'à son point le plus haut (du bas de la pile au haut du pylône), le Viaduc de Millau est d'une hauteur supérieure à celle de la Tour Eiffel. Les piles sont en forme de V, et les pylônes en forme de V renversé, cette solution permettant de minimiser l'impact sur le paysage tout en offrant une résistance suffisante.

From left to right, from above to below:
Aerial view, perspective,
view along the bridge.
Right: Approach to the bridge.

Von links nach rechts, von oben nach unten:
Luftaufnahme, Perspektive,
Blick entlang der Brücke.
Rechts: Auffahrt zur Brücke.

De gauche à droite, du haut vers le bas:
Trois vues aériennes prises
sous différents angles.
À droite: Autoroute menant au viaduc.

PASSERELLE VALMY,

NANTERRE / LA DÉFENSE, FRANCE

DIETMAR FEICHTINGER ARCHITECTES

www.feichtingerarchitectes.com

Type: Pre-stressed asymmetric cable-stayed bridge, **Length:** 88 m, **Structural engineers:** Schlaich Bergermann und Partner, **Completion:** 2007, **Photos:** Courtesy of Dietmar Feichtinger Architectes, Paris.

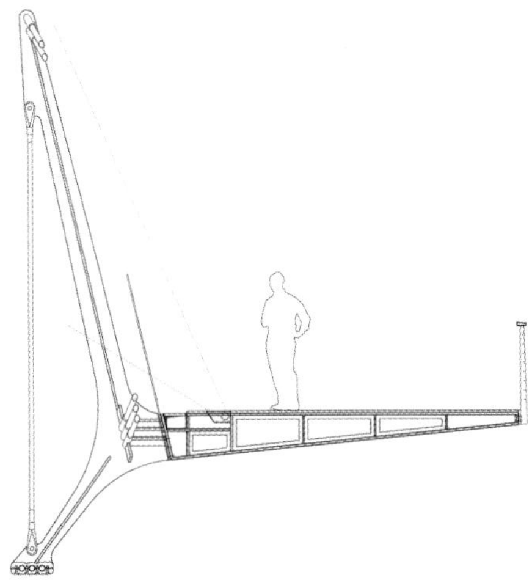

Left: General view. Links: Gesamtansicht. À gauche: Vue d'ensemble. | Right: Cross section. Rechts: Querschnitt. À droite: Vue en coupe.

The 88-meter long footbridge is part of a promenade between La Defense and Nanterre. Its layout responds to the geometry of the Société Générale office building. The structure is composed by supporting elements in the form of "vertebras". The "backbone" of the structure consists of a box girder made of welded steel sheets. The "vertebras" are linked with steel rods forming a cable-stayed structure for the vertical loads. Three cables following the outside edge of the circle situated under the deck level respond to torsion forces.

Die 88 Meter lange Promenade ist eine urbane Wegverbindung zwischen La Défense und Nanterre. Sie nimmt im Grundriss die Geometrie des Bürogebäudes der Société Génerale auf. Hauptmerkmal der Konstruktion sind die Stützelemente in Form von „Wirbeln", die die Promenade begleiten und rythmisieren. Das Rückgrat der Konstruktion bildet ein Hohlkastenträger aus verschweißten Stahlblechen. Die ausgreifenden „Wirbel" sind durch Zugstäbe miteinander verbunden, die vertikalen Kräfte werden über ein System von Kragträgern aufgenommen. Das dem Kreissegment folgende unter dem Brückendeck liegende Zugband ist mit den Wirbeln verbunden und dient der Aufnahme der Torsion über das horizontale Kräftepaar.

La passerelle de quatre-vingt mètres de long relie La Défense à Nanterre en contournant la tour de la Société Générale. Elle se caractérise par une succession d'éléments porteurs dont la forme évoque des vertèbres. Des caissons en acier assurent la rigidité de l'ensemble. Les « vertèbres » sont reliées entre elles par des haubans pour les efforts verticaux. Trois « tendons » en acier parallèles au tablier reprennent les efforts de torsion.

From left to right, from above to below:
Vertebras and backbone, general view,
detail, bridge between high-rise buildings.
Right: Structural detail.

Von links nach rechts, von oben nach unten:
Verspannung, Gesamtansicht, Strukturdetail,
Brücke zwischen den Hochhäusern.
Rechts: Strukturdetail.

De gauche à droite, du haut vers le bas:
« Vertèbres » avec haubans et « tendons »,
vue de la passerelle contournant la tour,
détail, la passerelle entre les tours.
À droite: Détail.

PASSERELLE SIMONE-DE-BEAUVOIR,
PARIS, FRANCE

DIETMAR FEICHTINGER ARCHITECTES

www.feichtingerarchitectes.com

Type: Arched and suspension bridge, **Length:** 304 m, **Structural engineers:** RFR, **Completion:** 2006, **Photos:** David Boureau, Paris (84, 87), Jo Pesendorfer, Paris (85 a. l.), Marc Verhille (85 a. r.).

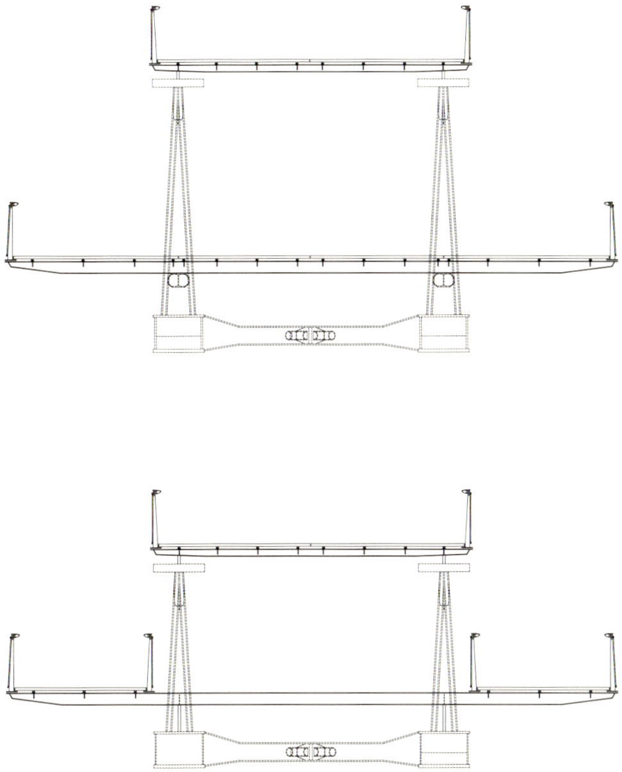

Left: View by night. Links: Nachtaufnahme. À gauche: Passerelle la nuit. | Right: Sections. Rechts: Schnitte. À droite: Vues en coupe.

The pedestrian bridge with a width of twelve meters and an overall length of 304 meters connects the French National Library and the Bercy Park. Two curves create the geometry of the bridge, whose soft shape contrasts with the surrounding buildings. The combination of the upgoing - compressed - arc and the inversed – tensioned – arc enable the free span of 194 meters. The overlay of the two curves creates a lens formed space in the middle of the bridge, a public area suspended over the water. The multiple crossing options following the curves of the structure lead the pedestrians close to the water or allow a panoramic view over Paris.

Die zwölf Meter breite und 304 Meter lange Fußgängerbrücke verbindet die französische Nationalbibliothek mit dem Park von Bercy. Zwei Kurven erzeugen die Geometrie der Brücke, deren weiche Form sich als Antwort auf das streng geometrische Umfeld versteht. Die Kombination eines Druckbogens und eines Zugbandes gewährleistet ein stabiles Tragwerk und ermöglicht die freie Spannweite von 194 Metern. Im zentralen Bereich der Brücke erzeugt die Überlagerung der Konstruktion ein Volumen in Form einer Linse und bildet einen öffentlichen Ort über dem Fluss. Durch die am Tragwerk entlang führenden multiplen Überqeuerungsmöglichkeiten wird die Nähe zum Wasser und der Bezug zu Paris hergestellt.

Longue de 304 mètres et large de douze, cette passerelle relie la Bibliothèque Nationale de France au parc de Bercy. Ses courbes s'inscrivent en contraste par rapport à la rigidité géométrique des bâtiments voisins. Un premier arc concave et un second convexe assurent la rigidité de l'ensemble sur une portée de 194 mètres. De leur intersection au milieu du fleuve résulte un espace couvert en forme de lentille. Les nombreux points de passage d'un niveau à l'autre permettent aux usagers tantôt d'être juste au-dessus de l'eau, tantôt d'apprécier une vue panoramique sur Paris.

From left to right, from above to below:
Crossing bridge decks, view with the Parc de Bercy,
general view with the Bibliotheque Nationale.
Right: The central area.

Von links nach rechts, von oben nach unten:
Sich kreuzende Brückendecks, Ansicht mit Parc de Bercy,
Gesamtansicht mit Bibliotheque Nationale.
Rechts: In der Linse der Brücke.

De gauche à droite, du haut vers le bas:
Double tablier, vue vers le parc de Bercy,
vue d'ensemble avec la BNF à l'arrière plan.
À droite: La « lentille ».

HUMBOLDTHAFEN CENTRAL STATION BRIDGE,
BERLIN, GERMANY

GMP - ARCHITEKTEN VON GERKAN, MARG UND PARTNER

www.gmp-architekten.de

Type: Steel concrete bridge with steel supports, **Length:** 250 m, **Concept:** Meinhard von Gerkan, Jürgen Hillmer und Jörg Schlaich, **Structural engineers:** Schlaich Bergermann und Partner, **Completion:** 2002, **Photos:** Jürgen Schmidt, Köln (89), Christine Maier (91, 92).

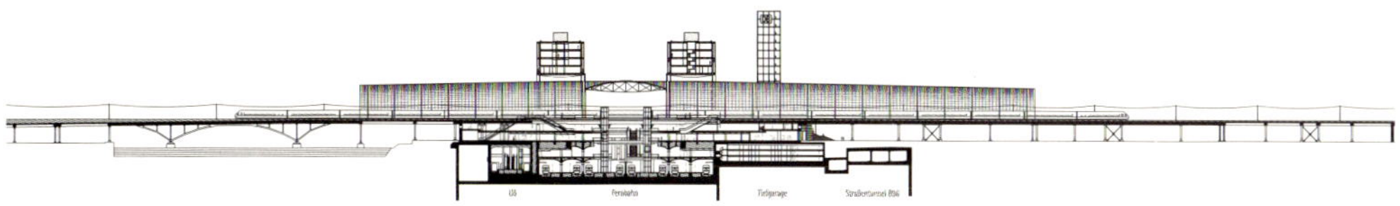

Left: Southern view of the bridge. Links: Südansicht der Brücke. À gauche: Pont vu du sud. | Right: Section. Rechts: Schnitt. À droite: Vue en coupe.

On the premises of the Berlin Central station, four railway and two city rail tracks had to be elevated at a height of 10 meters across one kilometer. For the highly visible bridge, a low construction height with an unobstructed view was desired. This was achieved by narrow support distances of up to 25 meters. With a span of up to 60 meters, the bridge is supported by steel pipe arches made of thick-walled seamless pipes. The special features of the 250-meter long bridge, which received the German bridge construction award in 2008, are the trumpet-shaped enlargements from East to West, as well as the cross sections of the bearing structures, which consist of arched cantilever arms and basket arches.

Im Bereich des Berliner Hauptbahnhofes mussten auf einem Kilometer vier Fernbahn- und zwei S-Bahngleise in 10 Meter Höhe überführt werden. Für die Brücke in exponierter Lage wurde eine geringe Bauhöhe mit möglichst freier Durchsicht gewünscht. Dies wurde mit engen Stützenabständen von bis zu 25 Metern erreicht. Die Brücke mit Spannweiten bis 60 Metern wird von Stahlrohrbögen aus dickwandigen, nahtlosen Rohren gestützt. Das Besondere an der 250 Meter langen Brücke, die 2008 mit dem Deutschen Brückenbaupreis ausgezeichnet wurde, sind die trompetenförmigen Aufweitungen von Ost nach West sowie die Querschnitte der Tragwerke, die sich aus Kragarmen mit bogenförmiger Untersicht und Korbbögen zusammensetzen.

L'aménagement de la gare centrale de Berlin a rendu nécessaire de construire six ponts de chemin de fer pour quatre grandes lignes et deux lignes de trains de banlieue. Le pont situé le plus près du centre ville devait être aussi bas et transparent que possible pour des raisons esthétiques. Les architectes ont résolu le problème avec un espacement latéral des piles de vingt-cinq mètres, pour une portée d'une de soixante. Les supports ont été réalisés avec des tubes d'acier à paroi épaisse d'un seul tenant. Parmi les caractéristiques de cet ouvrage de 250 mètres de long qui a reçu le Prix 2008 des ponts allemands, citons notamment le plan « en trompette » et la conception particulière des piles, composées de consoles formant des arcs en anse de panier.

From left to right, from above to below:
Under the bridge, structural detail, bridge with city rail.
Right: Transition to the central station.

Von links nach rechts, von oben nach unten:
Unter der Brücke, Strukturdetail, Brücke mit S-Bahn.
Rechts: Überleitung in den Hauptnahnhof.

De gauche à droite, du haut vers le bas:
Deux vues du pont par en dessous mettant
en évidence la structure des piles,
train de banlieue traversant le pont.
À droite: Vue du pont et de la
nouvelle gare centrale.

NORDPOL BRIDGE WESTPARK,
BOCHUM, GERMANY

HEGGER HEGGER SCHLEIFF – HHS PLANER + ARCHITEKTEN AG

www.hhs-architekten.de

Type: Cantilever and girder bridge, **Length:** 100 m, **Structural engineers:** Bollinger + Grohmann Ingenieure, **Light design:** Achim Wollscheid, Frankfurt / Main, **Completion:** 2000, **Photos:** Constantin Meyer, Cologne.

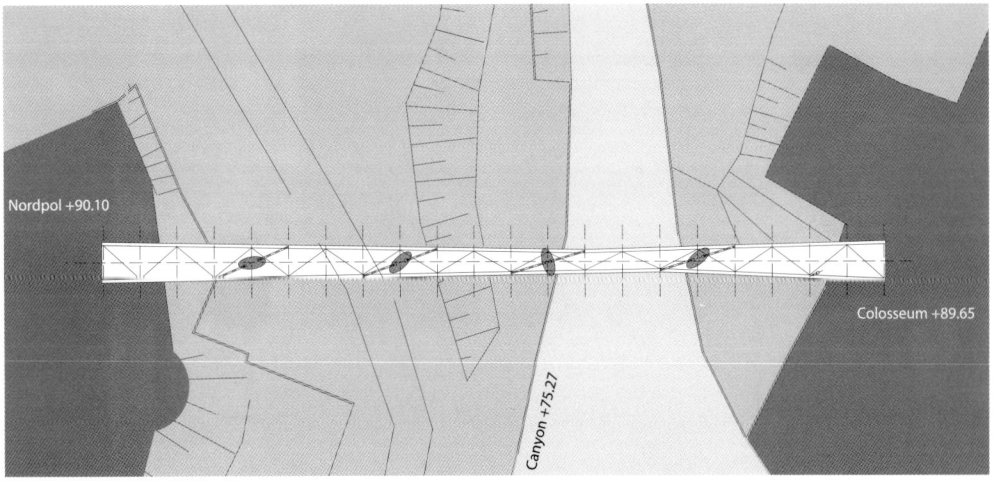

Left: Bridge head plateau 'Nordpol'. Links: Brückenkopf Plateau ‚Nordpol'. À gauche: Culée et passerelle. | Right: Site plan. Rechts: Lageplan. À droite: Plan de situation.

In the Westpark of Bochum, the Nordpol bridge links a former bridge abutment ('Nordpol') to a former blast furnace plateau ('Colosseum'). This makes the bridge a central element of the park premises, which were transformed into an industrial technology park during the land restoration of the former factory premises of the Krupp company. The polygonal ground plot and the gate-like V-supports resulted from the existing structures. The bridge is designed as a suspended lattice truss with a translucent walk-on finish. A light installation reacts to the dynamics of the bridge. The reflection effect is enhanced by the special disc structure of the rail bracing.

Im Westpark Bochums verbindet die Nordpolbrücke ein ehemaliges Brückenwiderlager (‚Nordpol') mit einem ehemaligen Hochofenplateau (‚Colosseum'). Sie ist damit ein Hauptelement des Rundgangkonzeptes im Park, der im Rahmen einer Rekultivierung des ehemaligen Fabrikgeländes der Firma Krupp in einen industriearchäologischen Park umgewandelt wurde. Der polygonale Grundriss und die torartig aufgestellten V-Stützen wurden aus der bestehenden Situation abgeleitet. Die Brücke ist als liegender Fachwerkträger mit durchscheinendem Gehbelag konzipiert. Ein Lichtkunstwerk reagiert auf die Dynamik auf der Brücke. Der Reflexionseffekt wird durch den besonderen Scheibenaufbau der Geländerausfachung verstärkt.

Lors de la reconversion des usines Krupp de Bochum en parc d'archéologie industrielle, une nouvelle passerelle a été construite entre une ancienne coulée de pont (« Pôle Nord ») et l'esplanade des hauts fourneaux (« Colisée »). Cet élément central du nouveau parc (notamment son plan polygonal et ses supports intermédiaires en forme de « V ») a été conçu en fonction du contexte de départ. Le tablier se compose d'une structure de poutrelles revêtue de grilles métalliques. Les plaques de verre du parapet créent un effet réfléchissant qui vient renforcer le dynamisme de l'ouvrage.

From left to right, from above to below:
Bottom view of the bridge across its entire length, side view,
Side view with V-supports, embedment of the
"Colosseum" bridge head plateau.
Right: Bridge at night.

Von links nach rechts, von oben nach unten:
Brückenuntersicht über die Gesamtlänge, Seitenansicht,
Seitenansicht mit V-Stützen, Einbettung des
Brückenkopfes Plateau ‚Colosseum'.
Rechts: Brückenlauf bei Nacht.

De gauche à droite, du haut vers le bas:
La passerelle vue d'en dessous, vue latérale, autre vue latérale
avec supports en V, vue dans l'axe en direction du « Colisée ».
À droite: Lla passerelle au crépuscule.

BLECHENSTEG,
COTTBUS, GERMANY

HENRY RIPKE ARCHITEKTEN

www.ripke-architekten.de

Type: Steel box girder, **Length:** 38.40 m, **Structural engineers:** VIC Brücken und Ingenieurbau GmbH, Potsdam, **Completion:** 2005, **Photos:** Courtesy of Henry Ripke Architekten, Berlin.

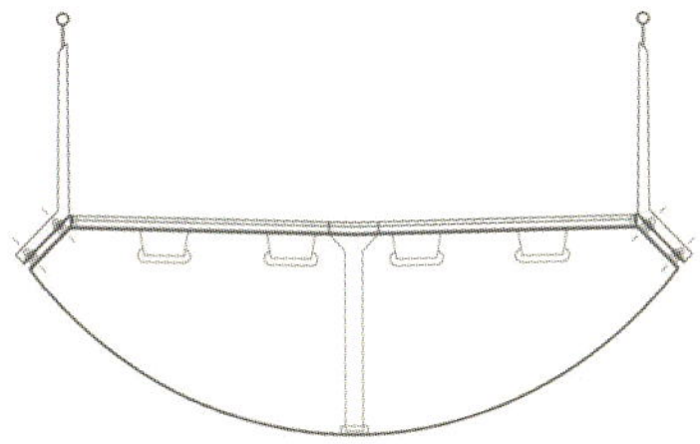

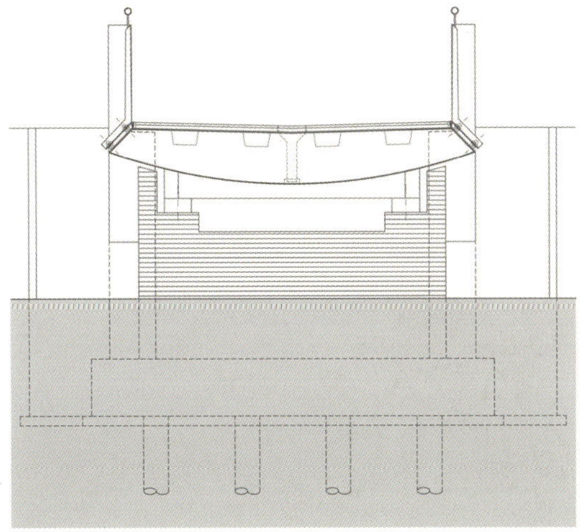

Left: Handrail detail. Links: Geländerdetail. À gauche: Gros plan sur la rambarde. | Right: Cross-section of the bearing and the center of the bridge. Rechts: Querschnitt am Widerlager und in Brückenmitte. À droite: Vue en coupe.

The design concept is based on the gentle fitting of the footbridge into the protected park landscape. The tension of the bridge results from the elegant crescent shape of the continuously changing box girder, which culminates in repetitious elements, the handrail posts. Bright reddish orange – in contrast to the anthracite of the remainder of the bridge – is only applied on the lateral sections of the handrail posts. This results in a special visual effect which seems as if the absorbed surface is moving along with the observer. A custom-made light band underscores the linearity of the bridge at night.

Die Gestaltungsidee basiert auf der behutsamen Einpassung des Steges in die denkmalgeschützte Parklandschaft. Die Spannung der Brücke entsteht aus der eleganten Sichelform des sich kontinuierlich verändernden Hohlkastens, dessen Abschluss aus einem repetierenden Element, dem Geländerstab, gebildet wird. Leuchtendes Rotorange – im Kontrast zum Anthrazit der übrigen Brücke – ist nur an den Seitenflächen der Geländerstäbe aufgetragen. In der perspektivischen Wahrnehmung entsteht dadurch der besondere Effekt, als wandere die rezipierte Fläche mit dem Betrachter mit. Ein eigens entwickeltes Lichtband im Handlauf unterstreicht nachts die Linearität der Brücke.

Cette passerelle a été conçue pour s'intégrer harmonieusement à un parc classé monument historique. Elle se caractérise par une forme incurvée très élégante et une rambarde en baguettes de bois qui sont peintes dans un ton rouge orangé lumineux et contrastent avec la couleur anthracite des autres éléments. Vue en perspective, la passerelle donne ainsi l'impression d'être en mouvement. La nuit, une bande lumineuse intégrée à la main courante souligne la linéarité de l'ouvrage.

From left to right, from above to below:
Handrail, perspective, view, view from below.
Right: View of bridge detail.

Von links nach rechts, von oben nach unten:
Geländer, Perspektive, Ansicht, Untersicht.
Rechts: Detailansicht der Brücke.

De gauche à droite, du haut vers le bas:
Rambarde et main courante, vue en perspective,
vue d'ensemble, la passerelle vue d'en dessous.
À droite: Vue de détail.

FIBERGLASS BRIDGE,
FRIEDBERG (HESSE), GERMANY

KNIPPERS HELBIG ADVANCED ENGINEERING

www.khing.de

Type: Frame bridge, **Length:** 27 m, **Structural engineers:** Knippers Helbig & König und Heunisch Planungsgesellschaft, **Completion:** 2008, **Photos:** Wilfried Dechau (100, 102 a. r., 102 b. r., 103), Fiberline Composites A/S, Denmark (102 a. l., 102 b. l.).

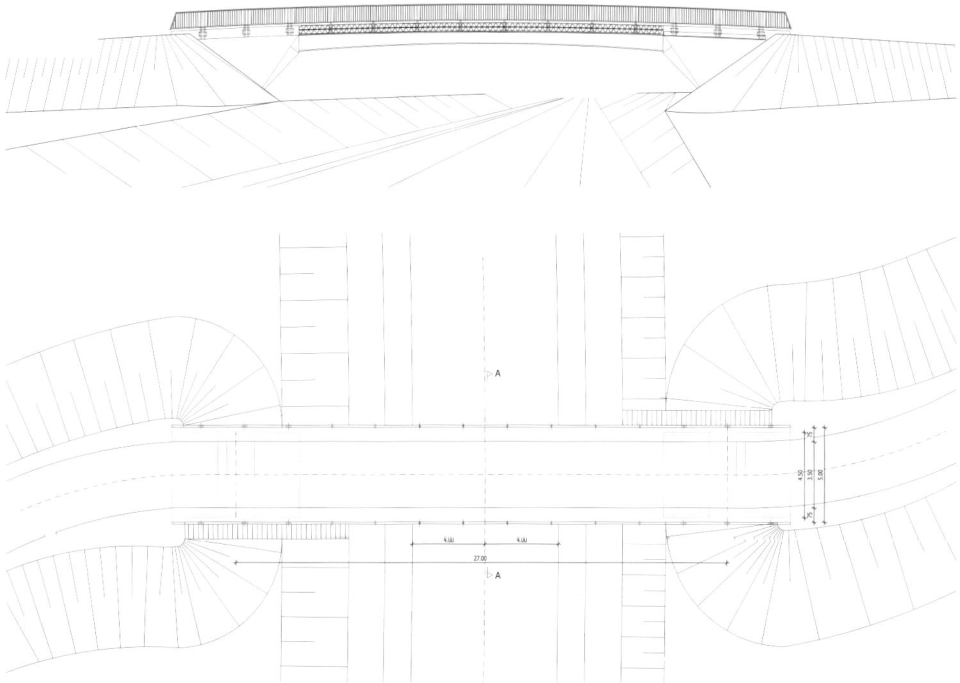

Left: View along deck. Links: Ansicht. À gauche: Vue du pont. | Right: Elevation and floor plan. Rechts: Ansicht und Grundriss. À droite: Élévation et plan.

For the first time in Germany, a street bridge made of fiberglass was created in Friedberg (Hesse) in July 2008. The fiberglass reinforced polymers (FRP) composite construction spans 27 meters across the Bundesstraße B 455 national highway. The new material not only has excellent mechanical properties, but it also offers low weight in addition to high durability. This is why it was possible to prefabricate the entire bridge inside a hall and to install it at its final location within a few hours. This new construction method considerably decreases the construction time and promises especially long durability.

In Friedberg (Hessen) wurde im Juli 2008 erstmals in Deutschland eine Straßenbrücke aus glasfaserverstärktem Kunststoff (GFK) realisiert. Das Bauwerk in Stahl-GFK-Verbundbauweise überspannt mit 27 Metern die Bundesstraße B 455. Der neue Werkstoff hat nicht nur hervorragende mechanische Eigenschaften, sondern neben seiner hohen Dauerhaftigkeit auch ein geringes Eigengewicht. Daher war es auch möglich, die Brücke in einer Halle komplett vorzufertigen und innerhalb weniger Stunden an ihrem Bestimmungsort zu installieren. Diese neue Brückenbauweise verkürzt die Bauzeit erheblich und verspricht eine besonders lange Lebensdauer.

En juillet 2008, un premier pont en acier et matière plastique renforcée à la fibre de verre a été inauguré en Allemagne à Friedberg, dans le land de Hesse. Il s'agit d'un ouvrage de vingt-sept mètres de long destiné à franchir le futur contournement du village par la route nationale B 455. Le matériau utilisé pour le tablier du pont se caractérise par ses excellentes propriétés mécaniques, sa longévité et son faible poids. Ces propriétés ont permis de le réaliser entièrement en usine et de le monter sur place en l'espace de quelques heures seulement. La nouvelle technologie est promise à un bel avenir du fait de sa rapidité d'exécution, alliée à une longévité exceptionnelle.

From left to right, from above to below:
Positioning of the bridge, bridge with surroundings,
bridge on the low platform trailer, parapet detail.
Right: Bridge with surroundings.

Von links nach rechts, von oben nach unten:
Einhub der Brücke, Brücke mit Umgebung,
Brücke auf dem Tieflader, Brückenlager.
Rechts: Brücke mit Umgebung.

De gauche à droite, du haut vers le bas:
Positionnement du pont préfabriqué,
le pont dans son environnement,
pont livré par un semi-remorque
surbaissé, gros plan sur la rambarde.
À droite: Le pont dans son environnement.

RECONSTRUCTION OF THE BERLIN BRIDGE,
HALLE/SAALE, GERMANY

INGENIEURBÜRO GRASSL GMBH

www.grassl-ing.de

Type: Cable-stayed bridge, **Length:** 171 m, **Design:** Uwe Graul, **Structural engineers:** Ingenieurbüro Grassl GmbH, **Completion:** 2006, **Photos:** Courtesy of Ingenieurbüro Grassl GmbH.

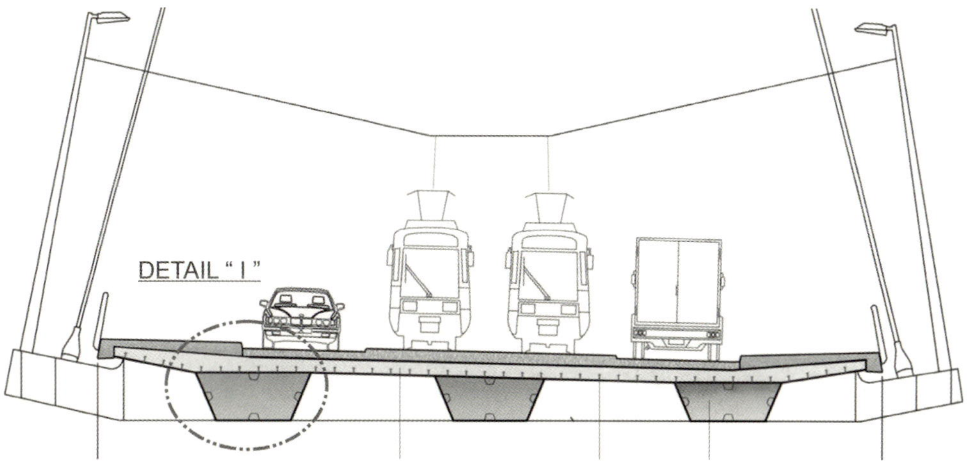

Left: Aerial view. Links: Luftaufnahme. À gauche: Vue aérienne. | Right: Section with stiffening girders. Rechts: Brückenquerschnitt und Versteifungsträger. À droite: Vue en coupe mettant en évidence les renforts composites.

The reconstruction of the Berlin Bridge involved Germany's first cable-stayed bridge with stiffening girders in a composite design of this size. The bridge, with its extremely arched layout, features motorways for individual and railway traffic. The bridge became an icon of the city within a very short period of time not only due to its impressive 73-meter high A pylon, but also due to its excellent integration into the cityscape and high design quality. The new entrance gate not only offers motorists improved traffic options, but also interesting views of the city.

Für den Neubau der Berliner Brücke wurde erstmalig in Deutschland eine Schrägseilbrücke mit Versteifungsträger in Verbundbauweise dieser Größe errichtet. Die Brücke mit ihrem stark gekrümmten Grundriss verfügt über Fahrbahnen für den Individual- und Schienenverkehr. Nicht nur durch ihren imposanten, 73 Meter hohen A-Pylon, sondern auch durch die sorgfältige Einbindung in das Stadtbild und den hohen gestalterischen Anspruch konnte die Brücke in kürzester Zeit zum Wahrzeichen der Stadt werden. Das neue Eingangstor gewährt den Verkehrsteilnehmern – abgesehen von einer verbesserten Verkehrsituation – auch interessante Ausblicke auf das Stadtgebiet.

Cet ouvrage est le plus grand pont haubané à renforts composites jamais construit en Allemagne. Le tablier, fortement incurvé, porte deux chaussées séparées pour voitures et camions et deux voies de tramway. Un pylône en forme de « A » haut de soixante-treize mètres, combiné aux efforts stylistiques réalisés afin d'intégrer le nouveau pont au paysage urbain, ont rapidement fait de l'ouvrage un symbole de Halle. Ce nouvel accès à la ville améliore la circulation locale, tout en offrant des vues panoramiques du fait de sa position dans un site dégagé.

From left to right, from above to below:
Top of the pylon, details, pylon.
Right: General view.

Von links nach rechts, von oben nach unten:
Spitze des Pylon, Seildetail,
Kabelverankerung, Pylon.
Rechts: Gesamtansicht.

De gauche à droite, du haut vers le bas:
Sommet du pylône, deux vues de l'ancrage
des haubans, vue générale du pylône.
À droite: Vue aérienne.

PEDESTRIAN AND BICYCLE BRIDGE NEUE BAHN STADT,
LEVERKUSEN, GERMANY

KNIGHT ARCHITECTS WITH KNIPPERS HELBIG ADVANCED ENGINEERING

www.knightarchitects.co.uk, www.khing.de

Type: Cantilever and girder bridge, **Length:** 100 m, **Completion:** 2011, **Renderings:** Courtesy of Knight Architets & Knippers Helbig.

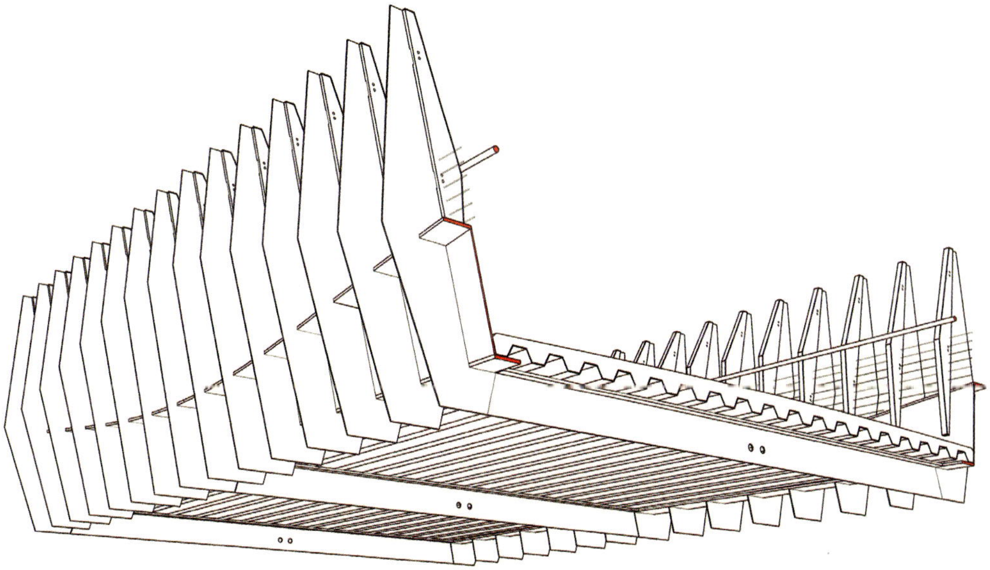

Left: Bridge deck. Links: Brückendeck. À gauche: Passants sur la passerelle. | Right: Construction principle. Rechts: Konstruktionsprinzip. À droite: Principe de la structure du tablier.

The pedestrian and bicycle path bridge links downtown Opladen with a revitalized railway area. The rigid steel sheets required for a steel construction were further developed into a distinctive slatted structure. The edges of the exterior steel slats and the interior wooden slats create a sculptural presence, with an open and transparent look in elevation and an enclosed view on axis. The deliberate omission of several individual slats also presents defined views of the city. The sides of the bridge beam are informed by the bending moment of each span, resulting in an elegant interaction between the wave forms of the beam and the slatted sides.

Die Fuss- und Radwegebrücken verbinden die Opladener Innenstadt mit einem revitalisierten Bahnareal. Die für eine Stahlkonstruktion notwendigen steifen Bleche werden hierbei zu einer prägnanten Lamellenstruktur weiterentwickelt. Die Kanten der äußeren Stahllamellen und hölzernen Innenlamellen erzeugen hierbei eine skulpturale Präsenz, offen und transparent in der Ansicht, umschlossen und gerichtet in der Flucht. Dabei werden durch gezieltes Auslassen einzelner Lamellen auch Sichtbezüge zur Stadt definiert. Die Seitenwangen der Brückenfelder sind affin zur Momentenbeanspruchung ausgeführt. So ergibt sich ein reizvoller Dialog der asynchronen Wellenbewegung von Lamelle und Seitenwange.

Cette passerelle pour piétons et cyclistes doit relier le centre ville d'Opladen aux alentours revitalisés de la gare de triage. Elle se compose de planches et de plaques d'acier usinées pour former des lamelles caractéristiques, d'une qualité toute sculpturale, qui forment un parapet interrompu transversalement mais continu dans le sens de la longueur. L'absence volontaire de lamelles à certains endroits offre aux passants des vues sur la ville. Cette conception particulière des lamelles du parapet s'allie harmonieusement à un tablier au profil qui ondule de manière asynchrone pour former un ensemble exceptionnellement dynamique.

From left to right, from above to below:
Visualisation, enlargement at the center
of the bridge, animation at night.
Right: Bridge balcony with staircase tower.

Von links nach rechts, von oben nach unten:
Animation, Aufweitung in Brückenmitte,
Animation bei Nacht.
Rechts: Brückenbalkon mit Treppenturm

De gauche à droite, du haut vers le bas:
Passerelle et train, élargissement du tablier
au milieu de la passerelle, passerelle la nuit.
À droite: Passerelle et cage d'escalier.

NEULANDBRÜCKE,
LEVERKUSEN, GERMANY

AGIRBAS / WIENSTROER, ARCHITEKTUR & STADTPLANUNG

www.aw-architektur.de

Type: Lattice bridge, **Length:** 165 m, **Structural engineers:** Arup GmbH, **Color design:** Friedrich Ernst von Garnier, Fürfeld, Dusseldorf, **Completion:** 2004, **Photos:** Nicola Roman Walbeck.

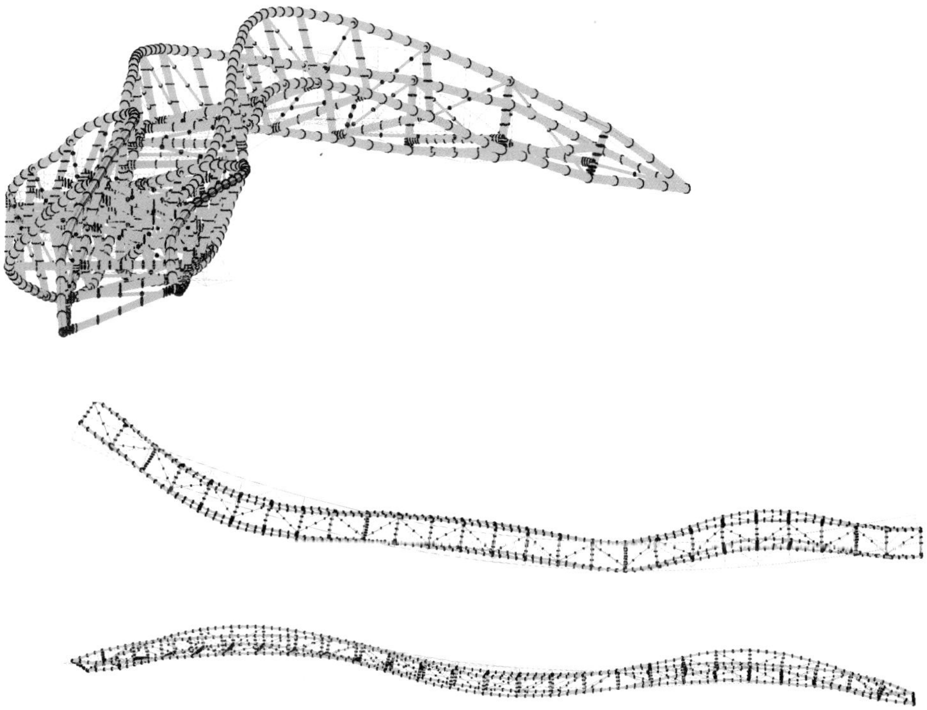

Left: Gate to the city. Links: Tor zur Stadt. À gauche: La passerelle côté ville. | Right: Diagrams of the distortion tests. Rechts: Diagramme der Verformungstests. À droite: Diagrammes des tests de distorsion.

The geometry and bearing structure of the Neuland bridge are defined by its surrounding setting. The bridge is part of the Bumerang trail and therefore has a bent layout, while the position of its supports is limited by the location of the Rheinallee and the media line extending in parallel. In addition, the heavy and special transport vehicles of the Rheinallee require a headroom of 4.75 meters. The large supporter distances are bridged by a steel square framed work. The resulting wave movement of this construction takes the various requirements into account, including the headroom and the unobstructed view of the parking area.

Geometrie und Tragkonstruktion der Neulandbrücke sind bestimmt durch die vorgefundenen Gegebenheiten: Die Brücke ist Bestandteil des Bumerangwegs und daher im Grundriss gekrümmt, die Stützstellung ist durch die Lage von Rheinallee und parallel liegender Medientrasse eingeschränkt, zusätzlich erfordern die auf der Rheinallee stattfindenden Schwer- und Spezialtransporte eine lichte Durchfahrtshöhe von 4,75 Metern. Die großen Stützabstände werden durch eine Fachwerkkonstruktion in Stahl überspannt. Die Wellenbewegung dieser Konstruktion als Ergebnis berücksichtigt die unterschiedlichsten Anforderungen, darunter die Durchfahrtshöhe an der Rheinallee und die freie Aussicht auf die Parkfläche.

La géométrie et la structure porteuse de cette passerelle sont le résultat direct de la topographie : la forme incurvée du tablier s'imposait du fait de l'intégration de l'ouvrage au « Bumerangweg » ; la disposition des piliers a été dictée par le tracé de la Rheinallee et de la Medienstrasse qui passent en dessous ; la surélévation au-dessus de la grande artère (4,75 mètres) était nécessaire du fait de la hauteur des camions qui y circulent. Une structure en poutrelles d'acier permet de compenser le grand écartement entre les piliers.

From left to right, from above to below:
Peripheral view, perspective view, linear and wave.
Right: View across the deck.

Von links nach rechts, von oben nach unten:
Seitenansicht, Perspektive, Linear und Welle.
Rechts: Blick über das Deck.

De gauche à droite, du haut vers le bas:
Vue latérale, perspective, linéarité et formes ondulées.
À droite: Vue du tablier.

LANGE BRÜCKE,
POTSDAM, GERMANY

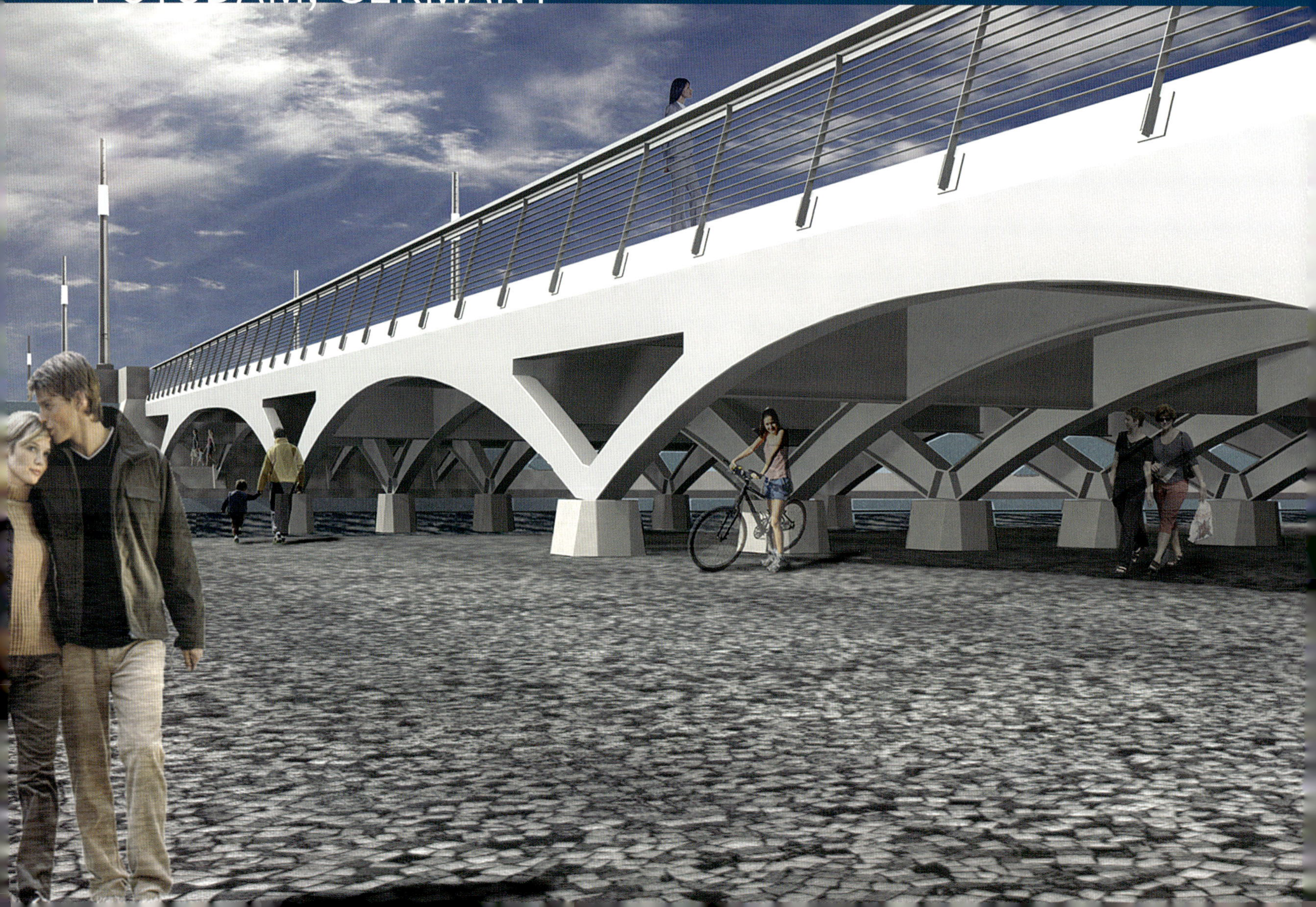

HENRY RIPKE ARCHITEKTEN, KLÄHNE INGENIEURE

www.ripke-architekten.de, www.kl-ing.de

Type: Multispan arch bridge, **Length:** 56.80 m + 68.20 m, **Structural engineers:** Klähne Beratende Ingenieure im Bauwesen GmbH, **Completion:** 2009 ongoing, **Photos:** Courtesy of Henry Ripke Architekten, Berlin (116, 118 o. l., 118 o. r., 119), Henry Ripke Architekten / Landeshauptstadt Potsdam (118).

Left: Perspective view Alte Fahrt. Links: Perspektivische Ansicht Alte Fahrt. À gauche: Vue en perspective des trois arches. | Right: View. Rechts: Ansicht. À droite: Vue en coupe.

The urban development project is defined by its crucial importance to the city of Potsdam. It is not only the oldest access to Potsdam but, in conjunction with the castle grounds, also constitutes the core of the urban development of the city. The new bridge is conceived as a classic urban bridge and creates an individual landmark without disrupting the context of the city with excessively constructive elements. Based on the basic concept of the "leap across the Havel", it consists of one large arch across the "Neue Fahrt" and three proportionally diminishing arches across the "Alte Fahrt". This way, both bridges look like a single structure reflecting on the previous structures.

Die städtebauliche Situation ist geprägt durch ihre herausragende Bedeutung für die Stadt Potsdam. Sie stellt nicht nur den ältesten Zugang nach Potsdam dar, sondern bildet mit dem Schlossareal den Nukleus der Stadtentwicklung Potsdams. Die neue Brücke wird als klassische Stadtbrücke interpretiert und setzt ein individuelles Zeichen, ohne durch übertriebene konstruktive Gesten den Kontext der Stadt zu stören. Ausgehend von dem Leitgedanken des „Sprungs über die Havel“ setzt sich die Bücke aus einem großen Bogen über die „Neue Fahrt“ und drei sich proportional verkleinernden über die „Alte Fahrt“ zusammen. Dadurch erscheinen die beiden Brücken wie ein Bauwerk, das die Vorgängerbauten reflektiert.

Il s'agit là du principal pont menant à Potsdam, situé à un emplacement utilisé depuis que la ville existe, à proximité du centre historique et de l'ancien palais royal. Dans ce contexte classique, les architectes ont privilégié des lignes individualisées mais empreintes d'une certaine retenue afin de ne pas perturber le paysage urbain. Le pont franchit la Havel au moyen d'une grande arche, continue sur l'île formée par la rivière, puis franchit le second bras au moyen de trois arches de tailles différentes. Le nouvel ouvrage d'art s'inscrit ainsi dans la tradition établie par le pont précédent.

hotel Mercure

From left to right, from above to below:
Bridge under construction, perspective view
Alte Fahrt, view from the south-east.
Right: Location of the completed project.

Von links nach rechts, von oben nach unten:
Brücke im Bau, Ansicht der Brücke über die Alte Fahrt,
Perspektive von Südosten.
Rechts: Lageplan des Endausbaus.

De gauche à droite, du haut vers le bas:
Vue en perspective durant les travaux,
pont en construction, vue aérienne.
À droite: Vue aérienne.

DRACHENBRÜCKE ACROSS THE GESSENTAL,
RONNEBURG, GERMANY

DIPL. ING. RICHARD J. DIETRICH
BÜRO FÜR INGENIEUR-ARCHITEKTUR

www.dietrich-ingenieur-architektur.de

Type: Stress ribbon bridge, **Length:** 232.5 m, **Stress analysis:** Köppl Ingenieure GmbH, Rosenheim, **Completion:** 2006, **Photos:** Richard J. Dietrich, Bergwiesen (120, 122 a. r., 122 b. l., 122 b. r., 123), BUGA Gera / Ronneburg 2007 GmbH (122 a. l.).

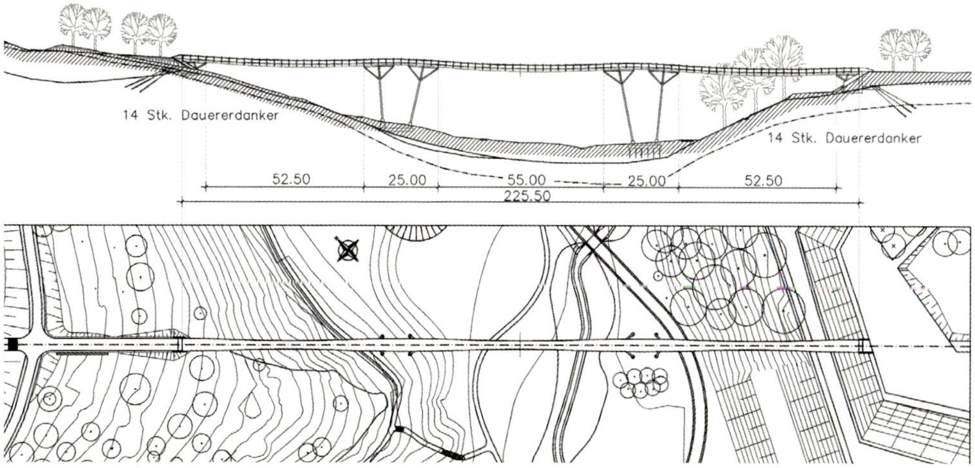

Left: Perspective. Links: Perspektive. À gauche: Perspective. | Right: Elevation and plan. Rechts: Ansicht und Grundriss. À droite: Élévation et plan.

This stress ribbon bridge is one of the world's largest and most innovative wooden bridges. The bearing structure consists of a stack glued waisted glulam plank. With a thickness of only 50 centimeters, freely stretches wavelike across three 50-meter wide fields. The tension band is gently curved and diverted across the tree-like strutted steel pipe pillars. The slag is optimally constructed. However, to prevent the pathway from becoming too steep, the band is increasingly doubled towards the center of the field, which also reduces the vibration. The waisting of the band improves its use and strengthens it. The wooden band is permanently weatherproofed.

Diese Spannbandbrücke ist eine der längsten und innovativsten Holzbrücken der Welt. Das Tragwerk besteht aus einer blockverleimten, taillierten Brettschichtholztafel, die mit nur 50 Zentimetern Stärke wellenförmig über drei 50 Meter weite Felder frei gespannt ist. Über den baumartig verstrebten Stahlrohrpfeilern wird das Spannband sanft gekrümmt umgeleitet. Der Durchhang ist konstruktiv optimiert. Damit jedoch die Gehbahn nicht zu steil wird, ist das Band zur Feldmitte hin zunehmend aufgedoppelt, was auch zur Schwingungsdämpfung beiträgt. Die Taillierung des Bandes dient sowohl der Nutzung als auch der Versteifung. Das Holzband ist dauerhaft wettergeschützt.

Ce pont caténaire est un des ouvrages en bois les plus longs et les plus novateurs au niveau mondial. Le tablier, épais de seulement cinquante centimètres et composé de lamelles encollées, repose sur deux piles doubles en tubes d'acier avec renforts supérieurs. Il forme ainsi trois portées ondulantes d'une cinquantaine de mètres chacune. Les ondulations ont été spécialement étudiées, le tablier étant par ailleurs renforcé en son milieu afin d'éviter une pente trop importante, d'absorber les vibrations et d'accroître la rigidité de l'ensemble. Le bois utilisé est pourvu d'un revêtement lui permettant de résister durablement aux intempéries.

From left to right, from above to below:
Swaying bridge deck, view from below with context,
general view, integration of bridge in the path network
Right: Swaying pedestrian route.

Von links nach rechts, von oben nach unten:
Geschwungenes Brückendeck, Untersicht mit Landschaft,
Gesamtansicht, Einbindung der Brücke in das Wegenetz.
Rechts: Der schwungvolle Brückenweg.

De gauche à droite, du haut vers le bas:
Piétons sur le tablier ondulant, pont vu d'en dessous,
vue d'ensemble, pont intégré dans le réseau
de sentiers de promenade.
À droite: Tablier ondulant.

LUISENBRÜCKE, BRÜCKE KULTURUFER, FUSSGÄNGERBRÜCKE BERLINER PROMENADE

SAARBRÜCKEN, GERMANY

STEPHAN BRAUNFELS ARCHITEKTEN

www.braunfels-architekten.de

Type: Arched bridges, **Length:** 82 m, 68 m, 57 m, **Engineers:** Sailer Stepan & Partner, Ingenieurbüro Grassl, **Completion:** 2015, **Renderings:** Courtesy by Stephan Braunfels Architekten.

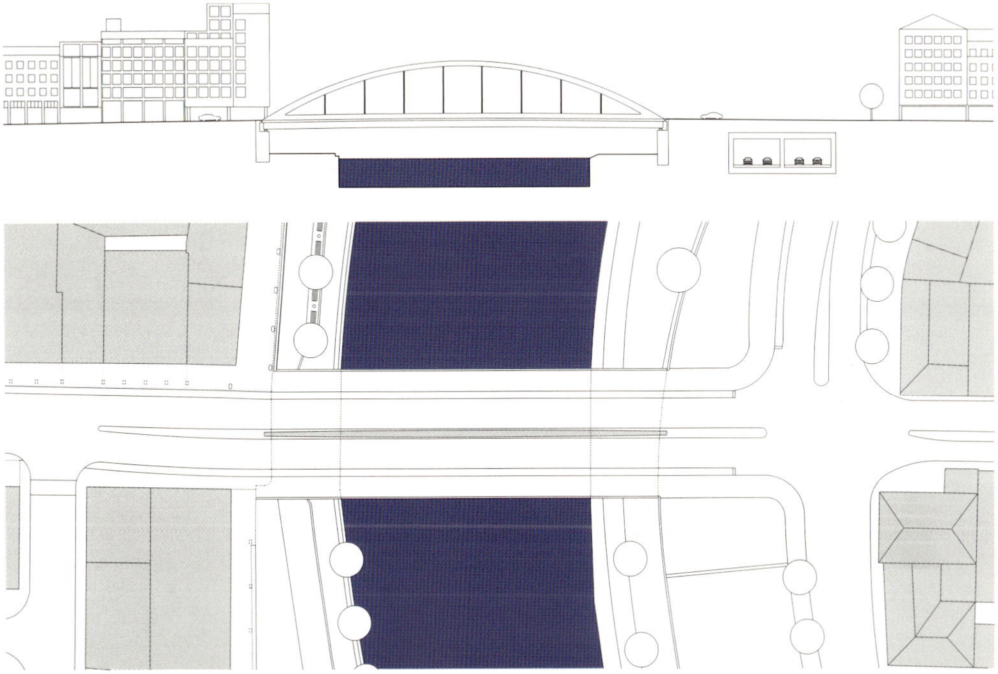

Left: Luisen bridge. Links: Luisenbrücke. À gauche: Luisenbrücke. | Right: Longitudinal section and floor plan. Rechts: Längsschnitt und Grundriss. À droite: Élévation et plan.

The inner city of Saarbrücken was in need of large-scale restructuring. The scope of this urban development measure included the new construction of three bridges over the Saar river. These included the Berliner Promenade pedestrian bridge, the reconstruction of the Luisen bridge, and the Kulturufer pedestrian bridge. The Luisen bridge is the longest of the three bridges. With an open span of 82 meters it extends completely across the river and the surrounding area. In consideration of the surrounding structures, it was designed as transparent as possible. The Berliner Promenade pedestrian bridge is designed as an arched bridge and also includes an curvature in its layout. It has a span of 68 meters.

Die Innenstadt Saarbrückens bedurfte einer groß angelegten Neustrukturierung. Im Zuge dieser Stadtentwicklungsmaßnahme wurde der Neubau dreier Brücken an der Saar geplant. Hierbei handelt es sich um die Fußgängerbrücke Berliner Promenade, den Neubau der Luisenbrücke und die Fußgängerbrücke Kulturufer. Die längste der drei Brücken ist die neue Luisenbrücke, welche mit einer Stützweite von 82 Metern den Fluss und das Vorland komplett überspannt. Sie wurde mit Rücksicht auf die angrenzende Bebauung so transparent wie möglich gestaltet. Die Fußgängerbrücke Berliner Promenade ist als Bogenbrücke angelegt und weist auch im Grundriss eine Krümmung auf. Sie erreicht eine Spannweite von 68 Metern.

Trois ponts ont été construits sur la Sarre dans le cadre de la restructuration du centre ville de Sarrebruck : passerelle de la Berliner Promenade, nouveau Luisenbrücke et passerelle du Kulturufer. Avec une longueur totale de quatre-vingt-deux mètres, le Luisenbrücke est le plus grand des trois ouvrages. Les architectes l'ont doté d'une silhouette transparente afin de minimiser l'impact sur l'espace urbain avoisinant. Quant à la passerelle de la Berliner Promenade, longue de soixante-huit mètres, elle se compose d'un tablier horizontal supporté par une arche dont le plan va en s'évasant pour accueillir des escaliers.

From left to right, from above to below:
Berliner Promenade bridge,
Luisen bridge, Kulturufer bridge.
Right: Night view.

Von links nach rechts, von oben nach unten:
Brücke Berliner Promenade,
Luisenbrücke, Brücke Kulturufer.
Rechts: Nachtansicht.

De gauche à droite, du haut vers le bas:
Vues de la passerelle de la Berliner Promenade,
Luisenbrücke, vue de la passerelle du Kulturufer.
À droite: Passerelle la nuit.

RÜGENBRÜCKE,
STRALSUND, GERMANY

ANDRÉ KEIPKE, KARL KLEINHANSS

www.architekt-keipke.de

Type: Cable-stayed bridge, **Length:** 2,830 m, **Design:** DEGES Berlin, **Planning partners:** Ingenieurgemeinschaft EHS-VCE, **Completion:** 2007, **Photos:** DEGES Berlin / R. Legrand.

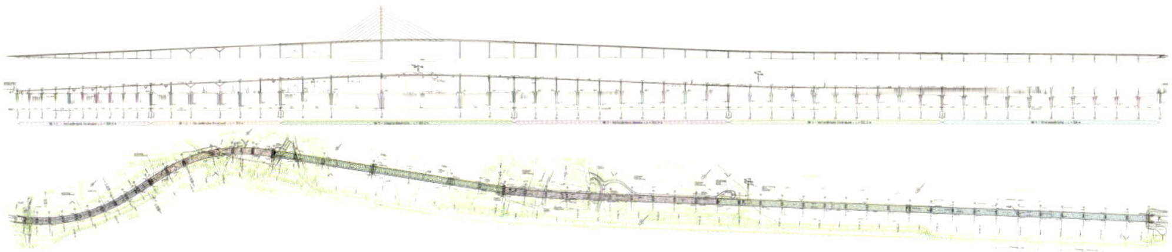

Left: Foot of the steel pylon. Links: Fußpunkt des Stahlpylons. À gauche: Pylône, haubans et tablier. | Right: General view. Rechts: Gesamtansicht. À droite: Élévation et plan.

The cable-stayed bridge across the Ziegelgraben with its 128 meter-high steel pylon constitutes the "New Gateway to Rügen" and is already considered to be the new landmark of the ancient Hanseatic city of Stralsund. The planners of the bridge paid particular attention to preserving the landscape and the medieval townscape of Stralsund by applying a light, visually permeable design. For the visually appealing cable-stayed bridge, which resembles a sailing ship, a split pylon with a slim and distinguished steel upper part was found to be the ideal solution. This dissolves the large overall structure and prevents a bulky appearance.

Die Schrägseilbrücke über den Ziegelgraben bildet mit dem 128 Meter hohen Stahlpylon das „Neue Tor nach Rügen" und gilt bereits als neues Wahrzeichen der alten Hansestadt Stralsund. Bei der Planung der Brücke wurde besonderer Wert auf die Erhaltung der Landschaft und des mittelalterlichen Stadtbildes von Stralsund gelegt und ein leichter, optisch durchlässiger Brückenzug entwickelt. Für die ansprechende, einem Segelschiff nachempfundene Schrägseilbrücke erwies sich ein zweigeteilter Pylon mit einem ebenso schlanken wie markanten Oberteil aus Stahl als die überzeugende Lösung. Auf diese Weise wird die Großstruktur aufgelöst und der Eindruck von Massigkeit vermieden.

Ce pont haubané qui évoque un grand voilier, d'ores et déjà emblématique de la ville hanséatique de Stralsund, assure la liaison entre le continent et l'île de Rügen. Les architectes l'ont doté d'une silhouette légère et transparente afin de mieux l'intégrer au contexte de la ville médiévale toute proche. Les haubans sont fixés à un double pylône de 128 mètres de haut dont la partie supérieure, d'une forme caractéristique et particulièrement élancée, a été réalisée en acier. Cette solution a permis de construire un ouvrage d'aspect svelte en dépit de ses dimensions imposantes.

B CV 6033

From left to right, from above to below:
View from the pylon, roadway, bottom cable connection,
Swaying deck from below.
Right: Stralsund Vorland bridge and Ziegelgraben bridge.

Von links nach rechts, von oben nach unten:
Blick vom Pylon, Fahrbahn, unterer Seilanschluss,
Geschwungenes Deck von unten..
Rechts: Vorlandbrücke Stralsund und Ziegelgrabenbrücke.

De gauche à droite, du haut vers le bas:
Le pont vu du haut du pylône, chaussée sur
le tablier, gros plan sur la fixation d'un hauban,
le pont vu d'en dessous à partir de Stralsund.
À droite: Vue d'ensemble.

NEUE MESSE STUTTGART,
STUTTGART, GERMANY

WULF & PARTNER, FREIE ARCHITEKTEN BDA

www.wulf-partner.de

Type: Lattice bridge, **Length:** 300 m, **Structural engineers:** Leonhardt, Andrä und Partner Beratende Ingenieure VBI GmbH, Stuttgart, **Landscape architects:** Adler+Olesch Freie Landschaftsplaner, **Completion:** 2007, **Photos:** Chris van Uffelen.

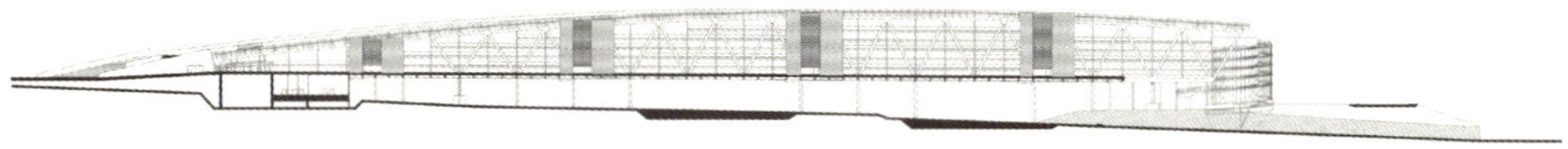

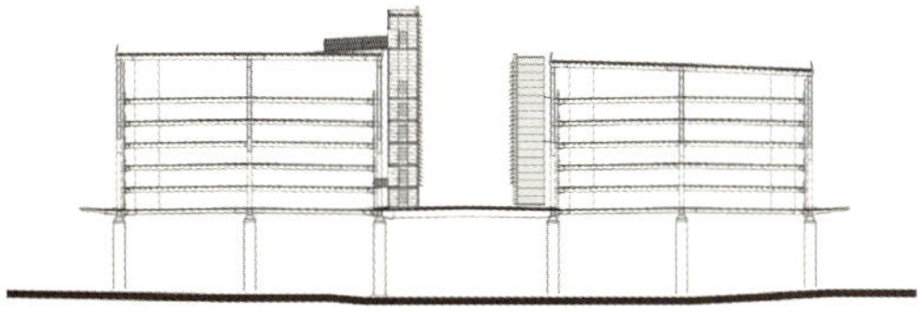

Left: : General view. Links: Gesamtansicht. À gauche: Vue d'ensemble. | Right: Sections. Rechts: Querschnitte. À droite: Deux vues en coupe.

This car park bridge across the A 8 highway is part of the Neue Messe Stuttgart (New Stuttgart fairgrounds) project, which combines the principles of minimal landscape consumption, short channels and an innovative appearance. The fairground axis was created with an East-West alignment parallel to the airport. This resulted in a connected construction island consisting of the airport and the fairgrounds. In addition, the car park with approximately 4,000 parking spaces was built as a bridge across the A8. The uniqueness of the overall project is primarily derived from the inclusion of the fair concept topography. The car park, which stretches as a "green bridge" across the highway, is the distinguishing feature of this design.

Diese Parkhausbrücke über die A 8 ist Teil des Projekts Neue Messe Stuttgart, welches die Maximen minimalen Landschaftsverbrauchs, kurzer Wege und eines innovativen Erscheinungsbildes in sich vereint. Die Messeachse wurde in Ost-West-Ausrichtung parallel zum Flughafen angelegt. Dadurch werden Flughafen und Messe zu einer zusammenhängenden Bebauungsinsel. Darüber hinaus wurde das Parkhaus mit ca. 4.000 Stellplätzen als Brücke über der A 8 gebaut. Die Unverwechselbarkeit der Gesamtanlage entsteht in erster Linie durch Einbeziehung der Topografie in das Messekonzept. Das Parkhaus, das sich als „Grünbrücke" über die Autobahn schwingt, ist das besondere Kennzeichen dieses Entwurfs.

Un parking a été construit au-dessus d'une autoroute à Stuttgart dans le cadre des travaux d'extension du parc des expositions. Le projet répond à trois idées principales : impact minimum sur le paysage, optimisation des distances et aspect futuriste. Les bâtiments du nouveau parc des expositions sont orientés est/ouest et parallèles à l'aéroport, de sorte que les deux équipements d'infrastructure forment un ensemble architectural cohérent. Le parking de quatre mille places a été conçu en fonction de la topographie locale. Il s'agit d'un « pont vert » qui traverse l'autoroute et sert d'image de marque au nouveau parc des expositions.

BOSCH
Stuttgart-Möhringen
Stuttgart-Degerloch
2000 m

View from adjacent bridge. Blick von benachbarter Brücke. Vue du pont sur l'autoroute.

BOSCH
Parkhaus

P20
A
B
C
F
P21

From left to right, from above to below:
Piles and truss, view between the
two bodies, detail, side elevation.
Right: General view.

Von links nach rechts, von oben nach unten:
Pfeiler und Fachwerk, Blick zwischen die
beiden Baukörper, Detail, Seitenansicht.
Rechts: Gesamtansicht.

De gauche à droite, du haut vers le bas:
Autoroute et piles du pont, espace entre
les deux bâtiments parallèles, détail,
autre vue de l'autoroute et du pont.
À droite: Vue d'ensemble.

DREILÄNDERBRÜCKE,

WEIL AM RHEIN / GERMANY, HUNINGUE / FRANCE

DIETMAR FEICHTINGER ARCHITECTES

www.feichtingerarchitectes.com

Type: Arched bridge, **Length:** 346 m, **Structural engineers:** LAP, **Completion:** 2006, **Photos:** A. Caste (138, 140 a., 141), D. Boreau (140 b. l., 140 b. r.).

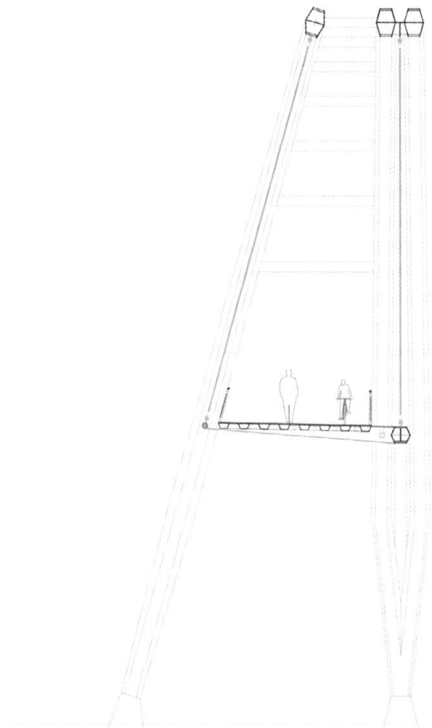

Left: General view. Links: Gesamtansicht. À gauche: Vue d'ensemble. | Right: Section. Rechts: Section. À droite: Vue en coupe.

The tripartite border bridge extends between Germany and France, constituting a symbol of the close relation between the two countries. The extremely shallow arch resulting from the low rise provides tension and elegance to the bridge. Its asymmetrical supporting structure is based on two arches, one of which is inclined and leaning on the other. This keeps the views of both river banks intact. The bridge bearings are located on the banks, keeping the river bed largely untouched. The bearing structure of the bridge deck is also asymmetrical. The sparsely applied illumination elements are integrated in the hand rail and the base points of the arch.

Die Dreiländerbrücke spannt einen Bogen zwischen Deutschland und Frankreich und ist ein Symbol für die enge Verbindung der Länder. Der durch seinen geringen Stich extrem flache Bogen verleiht der Brücke Spannung und Eleganz. Ihre asymmetrische Tragkonstruktion ergibt sich aus zwei Bögen, von denen sich einer neigt und an den anderen lehnt. Hierdurch bleibt die Sichtbeziehung der beiden Uferseiten erhalten. Die Auflager befinden sich im Uferbereich, sodass das Flussbett weitgehend unberührt bleibt. Die Tragstruktur des Brückendecks ist ebenfalls asymmetrisch ausgebildet. Die zurückhaltend eingesetzten Beleuchtungselemente sind in den Handlauf und in die Fußpunkte des Bogens integriert.

Cette passerelle au-dessus du Rhin symbolise l'union étroite entre la France et l'Allemagne. Son arche extrêmement basse lui confère une élégance particulière. Il s'agit d'une construction asymétrique dans le sens longitudinal, l'arche étant formée d'un arceau simple adossé à un arceau double. Ce type de construction présente l'avantage de laisser la vue libre d'une rive à l'autre. Les points d'appui se trouvent près des berges, ce qui a permis de minimiser l'impact sur le lit du fleuve. Un système d'éclairage discret a été intégré à la rambarde et au tablier.

From left to right, from above to below:
Structure, view towards Huningue, detail stairs.
Right: Night view.

Von links nach rechts, von oben nach unten:
Konstruktion, Blick Richtung Huningue,
Detail Stiegenaufgang.
Rechts: Nachtaufnahme.

De gauche à droite, du haut vers le bas:
Gros plan sur un point d'appui, vue en
direction de Huningue, détail de l'escalier.
À droite: La passerelle la nuit.

KATEHAKI PEDESTRIAN BRIDGE,
ATHENS, GREECE

SANTIAGO CALATRAVA ARCHITECT & ENGINEER

www.calatrava.com

Type: Cable-stayed bridge, **Length:** 93.7 m, **Completion:** 2004, **Photos:** Alan Karchmer / Architectural Photographer, **Plans:** Santiago Calatrava.

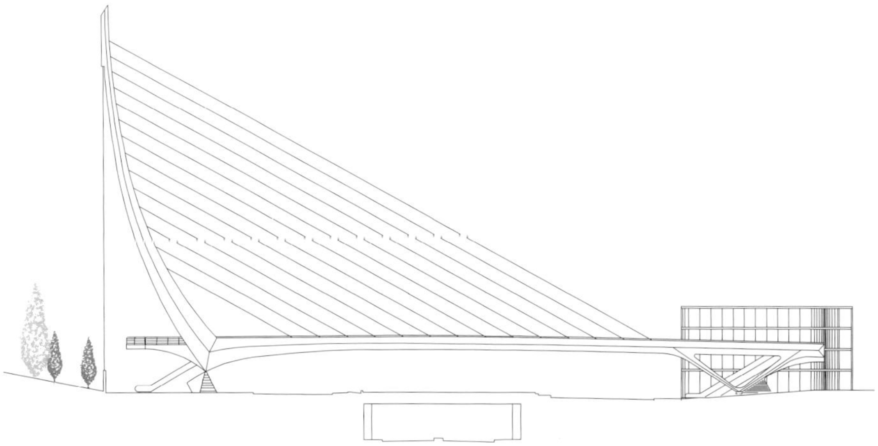

Left: Bridge with harp-like stays. Links: Brücke mit harfenartiger Verspannung. À gauche: Haubans évoquant une harpe. | Right: Elevation. Rechts: Ansicht. À droite: Élévation.

This pedestrian bridge crosses over Mesogion Avenue in Athens, connecting both sides of this street and the subway station. The footbridge is designed to be an integral part of the station as well as a landmark for the area. Two stairways, which are complemented by two escalators and two elevators provide access to the footbridge. The structure is a single-span stayed bridge with inclined pylon and exterior counterstays. However, this general typology is brought to its extreme by locating the deck's stays in a single lateral plane and by endowing the pylon with a variably curved shape, with the pylon's tieback located in a vertical plane of its exterior.

Die Fußgängerbrücke quert die Mesogion Avenue in Athen und verbindet beide Straßenseiten mit der U-Bahn-Station. Konzipiert ist sie sowohl als fester Bestandteil der U-Bahn-Station als auch als Wahrzeichen des Stadtviertels. Zwei Treppen, die durch zwei Rolltreppen und zwei Fahrstühle ergänzt werden, dienen als Zugang zur Brücke. Die allgemeine Struktur der Brücke – eine Schrägseilbrücke mit einzelnem betonummantelten Pylon und außenliegenden Gegenstreben – wurde hier ins Extrem getrieben, indem die Fahrbahnverstrebungen in einer einzigen seitlichen Ebene angebracht wurden und durch die ungleichmäßig geschwungene Form des Pylon mit seinen Gegenstreben an einer vertikalen Ebene seiner Außenseite.

Cette passerelle permet de traverser Mesogion Avenue pour accéder à une station de métro. Elle a été conçue en même temps que la station elle-même afin de doter le quartier d'une image forte. À chaque extrémité, l'accès se fait par un escalier, un ascenseur et un escalator. Il s'agit d'un ouvrage haubané à pylône unique et renforts extérieurs, ce type de construction étant ici poussé à l'extrême puisque le pylône, en position latérale, adopte une forme incurvée et se trouve uniquement retenu par un double câble vertical.

Athina
Athina
Kareas
Kifisias Av.
ΑΤΤΙΚΗ ΟΔΟΣ

From left to right, from above to below:
Detail of pylon, general view, bridge approach, underside of bridge.
Right: Detail platform and pylon.

Von links nach rechts, von oben nach unten:
Detail des Pylons, Gesamtansicht, Brückenaufgang, Unterseite.
Rechts: Detail Brückenplattform und Pylon.

De gauche à droite, du haut vers le bas:
Détail du pylône, vue d'ensemble, accès à la passerelle, avenue et tablier.
À droite: Détail de la plate-forme et du pylône.

THREE FOOTBRIDGES,
REYKJAVÍK, ICELAND

STUDIO GRANDA

www.studiogranda.is

Type: Post-tensioned concrete deck, **Length:** 150 m / 75 m / 85 m, **Structural engineers:** Línuhönnun consulting engineers, **Landscape architects:** Landslag Landscape architects, **Completion:** 2006, **Photos:** Sigurgeir Sigurjónsson, Reykjavík.

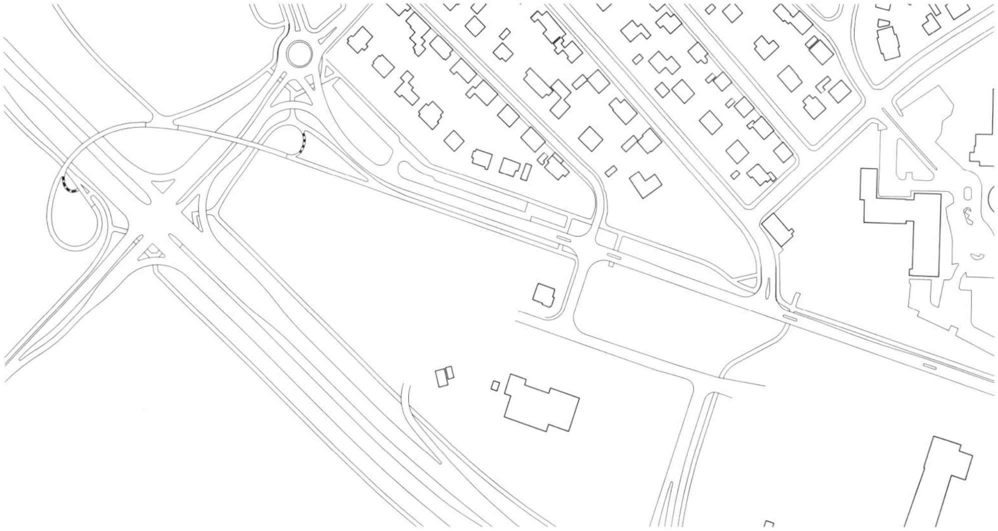

Left: Cross-point. Links: Kreuzungspunkt. À gauche: La passerelle vue d'en dessous. | Right: Site plan. Rechts: Lageplan. À droite: Plan de situation.

The construction of a new six-lane urban highway next to the domestic airport in the heart of Reykjavík required a new system of pedestrian routes including three footbridges. The design consists of a circular core of post-tensioned cables surmounted by the thinnest possible concrete deck. The structure rests on concrete filled stainless steel columns and minimal concrete landings. This arrangement allows the bridges to be curved in plan and section and tailored to each location without significant variation in the fundamental design. Expansion joints are only necessary at the landings as the columns and deck are sufficiently flexible to accept thermal movements.

Der Bau einer neuen Autobahn neben dem Inlandsflughafen Reykjaviks verlangte nach einem neuen System von Fußgängerwegen und drei dazugehörigen Brücken. Der Entwurf umfasst einen kreisförmigen Kernbereich nachgespannter Kabel, welcher mit einem besonders dünnen Betonbelag bedeckt ist. Getragen wird die Struktur von betongefüllten Edelstahlsäulen und Podesten mit geringer Höhe. Diese einfache Anordnung ermöglicht eine Krümmung der Brücke in Auf- und Querschnitt sowie eine Anpassung an jede beliebige Umgebung, ohne den Basisentwurf zu verändern. Dehnungsfugen sind ausschließlich an den Podesten nötig, da Fahrbahndecke und Säulen wärmebedingte Bewegungen ausreichend zulassen.

La construction d'une autoroute à deux fois trois voies desservant l'aéroport de Reykjavik a rendu nécessaire de construire également des passerelles pour les piétons et cyclistes habitant les quartiers voisins. Il s'agit d'ouvrages en béton précontraint composés de câbles et d'un tablier aussi fin que possible, l'ensemble reposant à la fois sur des culées elles aussi minimales, et sur des piles en béton armé intégrant de l'acier inox. Cette conception a permis de réaliser des courbes, tant sur le plan horizontal que vertical, de manière à optimiser l'adaptation de l'ouvrage à la topographie. Les joints de dilatation ne sont nécessaires qu'au niveau des culées car les piles et le tablier sont suffisamment flexibles pour s'adapter aux variations de température.

From left to right, from above to below:
Aerial view, parallel sweep, support detail, railing.
Right: Swinging transition.

Von links nach rechts, von oben nach unten:
Luftaufnahme, Parallelschwung, Stützendetail, Geländer.
Rechts: Schwungvoller Übergang.

De gauche à droite, du haut vers le bas:
Vue aérienne, passerelle double,
vue des piles, rambarde.

YAMUNA BRIDGE,
WAZIRABAD, NEW DELHI, INDIA

SCHLAICH BERGERMANN UND PARTNER STUTTGART-BERLIN-NEW YORK

www.sbp.de

Type: Cable-stayed bridge, **Length:** 575 m, **Architectural advise:** Ratan J. Batliboi, **Foundation design:** CONSTRUMA Constultancy PVT. LTD., **Completion:** 2015, **Renderings:** Courtesy of Schlaich Bergermann und Partner.

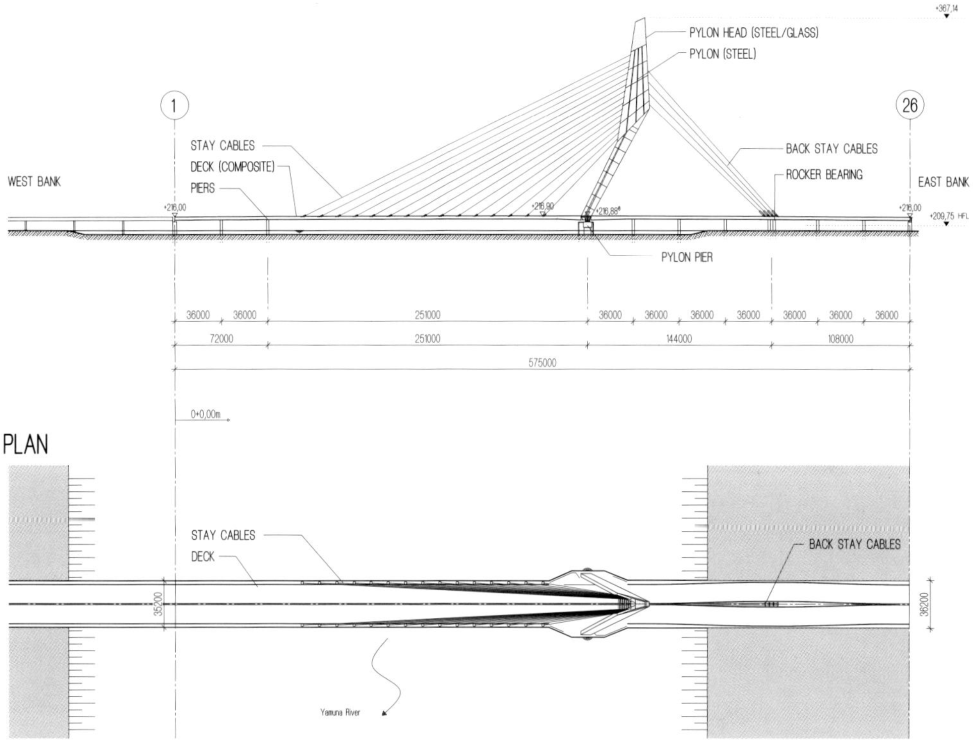

Left: "Tattoo bridge". Links: Brücke mit „Tätowierung". À gauche: Vue d'ensemble avec « tatouage ». | Right: Elevation and plan. Rechts: Ansicht und Grundriss. À droite: Élévation et plan.

With a harp-shaped cable-stay design, the monumental Signature Bridge crosses the Yamuna river with eight lanes. The dynamical shaped, inclined pylon provides to a substantial extent the stress balance required for supporting the deck. It also contains lifts and stairs for reaching the top of the pylon, where a visitors' gallery is planned. The pylon features two inclined supports, which are connected to the deck and meet at half height. The upper section of the pylon contains the cable stay mountings and the visitors' gallery. The composite deck consists of two outer main girders, a middle girder and cross girders with 4.5 meter spacing and concrete slab on top.

Diese imposante Brücke in harfenförmiger Schrägseilkonstruktion führt über den Fluss Yamuna. Der dynamisch geformte, geneigte Pylon sorgt in entscheidendem Ausmaß für den Lastausgleich, um die Fahrbahnplatte zu tragen. Er beinhaltet Fahrstühle und Treppengänge, damit die geplante Aussichtsplattform erreicht werden kann. Der Pylon besitzt zwei geneigte Stützen, die mit der Fahrbahnplatte verbunden sind und auf halber Höhe zusammentreffen. Der obere Abschnitt des Pylons beherbergt die Kabelverankerungen und die Besucherplattform. Die Fahrbahnplatte besteht aus zwei äußeren Hauptträgern, einem mittleren Träger und Querträgern aus Stahl im Abstand von 4,5 Metern mit aufgelegten Betonfertigteilen.

Ce pont à huit voies sur la rivière Yamuna se caractérise par un pylône central haubané qui donne à l'ouvrage l'aspect d'une harpe gigantesque. Le pylône est en forme de boomerang et contient des ascenseurs menant à une plate-forme panoramique avec vue sur la ville de Delhi. Il se compose de deux jambes inclinées qui se rejoignent à mi-hauteur, tous les haubans étant fixés à la partie supérieure. Le tablier se compose pour sa part de trois poutres longitudinales reliées par des poutrelles transversales espacées de 4,50 mètres, ainsi que d'une chaussée en béton.

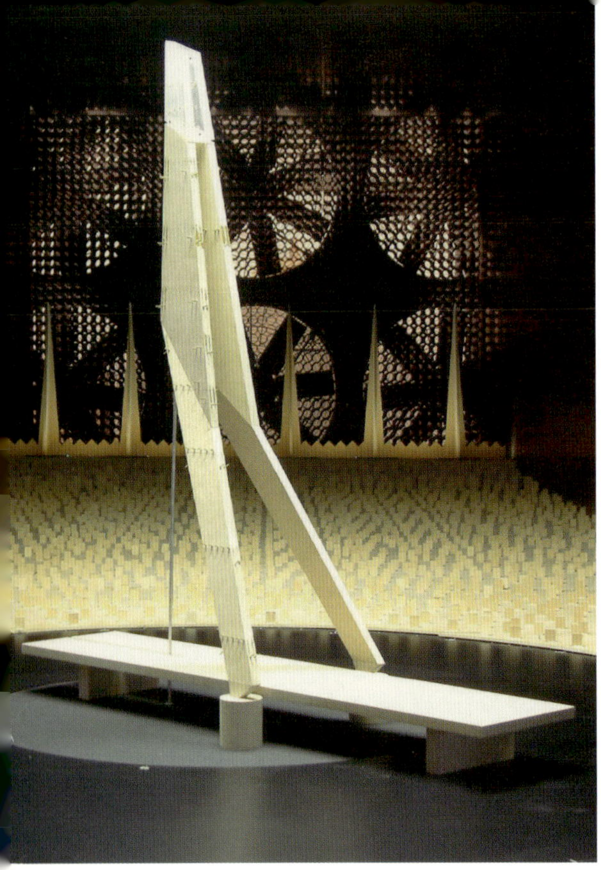

TO WAZIRABAD
PARKING
PARKING FOR BUS
PROPOSED BRIDGE
JETTY
JETTY

From left to right, from above to below:
Wind tunnel test, site plan,
pylon with illumination.
Right: Stay cables at pylon head.

Von links nach rechts, von oben nach unten:
Test im Windkanal, Situation,
Pylon mit Beleuchtung.
Rechts: Kabelverankerungen am Pylon.

De gauche à droite, du haut vers le bas:
Maquette du pont dans la soufflerie,
plan de situation, pylône avec éclairage.
À droite: Ancrage des haubans sur le pylône.

SEAN O'CASEY BRIDGE,
DUBLIN, IRELAND

BRIAN O'HALLORAN & ASSOCIATES ARCHITECTS

www.bohaa.ie

Type: Movable cable-stayed bridge, **Length:** 97.4 m, **Architectural project team:** Cyril O'Neil (architect), John Kelly (3d), Mark Carter (presentation), Deirdre Prince (technologist), **Structural engineers:** O'Connor Sutton Cronin (consulting engineers), **Consultant bridge engineers:** Movable Bridges - Richard Parrott, **Completion:** 2005, **Photos:** Barry Mason, Dublin (154, 156 a. l., 156 a. r., 156 b. l.), Gerry O'Leary, Dublin (156 b. r., 157).

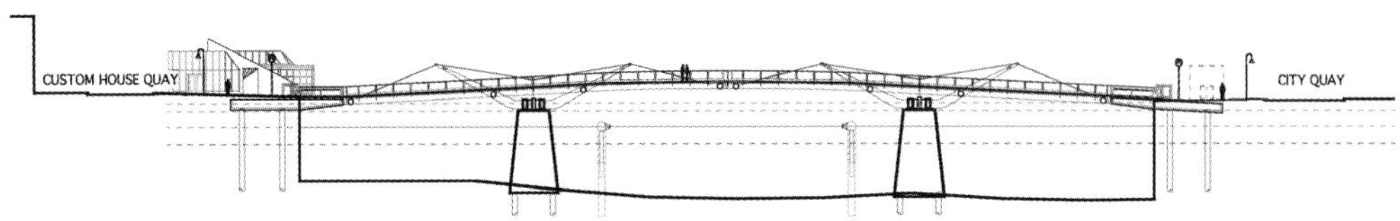

Left: Docker statue with bridge in the background. Links: Statue eines Hafenarbeiters mit der Brücke im Hintergrund. À gauche: Quai avec statue d'un docker et passerelle à l'arrière plan. | Right: Longitudinal section. Rechts: Längsschnitt. À droite: Vue en coupe.

The structural form of the two main bridge elements is based on a simple balanced cantilevered principle which provides stability while also providing the required width for passing vehicles. This resulted in a construction with a reduced structural mass and a dramatic and transparent silhouette. The two central opening sections have a total length of 44 meters and rotate at the same height to allow on opening width of 33 meters. Two profiled cantilever abutments complete the transition at the river banks. The suspended projection is supported by a frame comprised of four tapered steel sections, profiled to allow maximum load transfer.

Die Form der beiden Hauptelemente der Brücke basiert auf dem Prinzip des freien Vorbaus, welches eine stabile Struktur ermöglicht, die gleichzeitig die notwendige Durchfahrtsweite bei der Öffnung zulässt. Das Ergebnis ist eine an struktureller Masse reduzierte Konstruktion mit dramatischer und transparenter Form. Die zwei zentralen Öffnungsabschnitte haben eine Gesamtlänge von 44 Metern und drehen sich auf einer Höhe, um die nötige Durchfahrtsweite von 33 Metern zu ermöglichen. Zwei Widerlager komplettieren die Übergänge an den Uferseiten. Der freie Vorbau wird von einem Gerüst bestehend aus vier konisch zulaufenden Stahlsegmenten gestützt, die maximalen Ausgleich der Drucklast erlauben.

Les deux principaux éléments de cette passerelle mobile utilisent le principe du porte-à-faux, ce qui permet d'obtenir une structure rigide en dépit d'une silhouette filigrane et spectaculaire. Les éléments mobiles, d'une longueur totale d'environ quarante-quatre mètres, tournent sur un axe vertical de manière à ménager sur le fleuve une ouverture de trente-trois mètres de large. Ils se complètent sur chacun des quais par deux têtes de pont également en porte-à-faux. Quatre fourches en acier dont la position inclinée optimise la répartition des forces compensent le déséquilibre généré par le porte-à-faux.

From left to right, from above to below:
Bridge with Custom House in the background, bridge at night,
view out to sea, detail of the support cradle.
Right: Support cradle.

Von links nach rechts, von oben nach unten:
Brücke mit Custom House im Hintergrund, Brücke bei Nacht,
Blick seewärts, Detail des Stützgerüstes.
Rechts: Stützgerüst.

De gauche à droite, du haut vers le bas:
Passerelle avec la Douane à l'arrière plan,
passerelle la nuit, vue en direction
de la mer, fourche en acier.
À droite: Vue d'ensemble.

SPENCER DOCK BRIDGE,

DUBLIN, IRELAND

AMANDA LEVETE ARCHITECTS

www.amandaleveteararchitects.com

Type: Cantilever and girder bridge, **Length:** 40 m, **Structural engineers:** Arup, **Completion:** 2009, **Photos:** Gidon Fuehrer / Copyright Amanda Levete Architects.

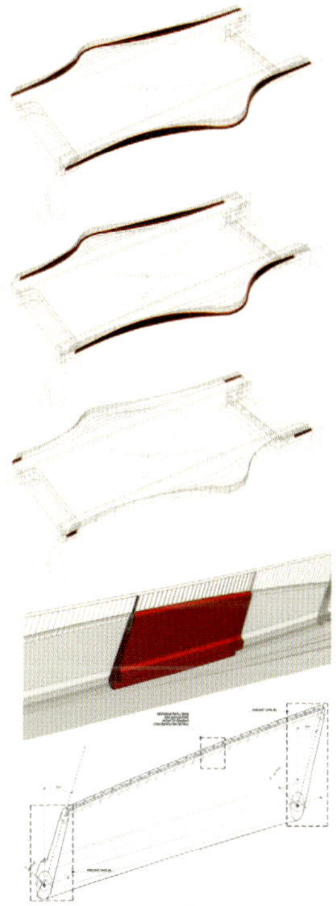

Left: View underneath the bridge. Links: Unterhalb der Brücke. À gauche: Le pont vu d'en dessous. | Right: Precast parapet. Rechts: Vorgefertigtes Geländer. À droite: Parapet préfabriqué.

The bridge has been constructed from an innovative combination of in situ and precast reinforced concrete. The form of the sculpturally formed bridge soffit was milled from high density expanded polystyrene and coated in resin to achieve a smooth surface finish suitable for maritime environments. Milled directly from 3-D parametric models, it provides a high degree of control over the geometry. This is the largest application of CNC cut polystyrene ever used. The concrete edges were precast and include recessed lighting and stainless steel balustrade infill bars. White limestone was used in the concrete mix to achieve a white finish, providing high visibility against the dark water.

Die Brücke wurde aus einer Kombination von Ortbeton und vorgefertigtem Stahlbeton errichtet. Die Form der skulpturalen Gewölbedecke wurde aus hochdichtem Polystyren gefräst und harzbeschichtet, um eine an maritime Verhältnisse angepasste, glatte Oberfläche zu erlangen. Für ein hohes Maß an Kontrolle über die Geometrie wurden beim Fräsvorgang parametrische 3D-Modelle verwendet. Dies ist die bislang größte Anwendung CNC-gefrästen Polystyrens. Die Betonkanten wurden vorgefertigt und beinhalten eingelassene Beleuchtungselemente sowie Geländer aus Edelstahl. Für die weiße Oberflächenfarbe mit ihrer guten Sichtbarkeit gegenüber dem dunklen Wasser wurde weißer Kalkstein im Betongemisch verwendet.

La construction de ce pont aux formes sculpturales s'est faite en utilisant à la fois du béton coulé sur place et des éléments préfabriqués en béton armé, ainsi que du polystyrène expansé revêtu de résine synthétique afin d'obtenir la surface lisse nécessaire à un environnement marin. L'usage de machines CNC taillant le polystyrène directement à partir de maquettes paramétriques en trois dimensions a permis de maîtriser parfaitement la géométrie du pont. C'est d'ailleurs le plus grand ouvrage jamais réalisé en utilisant cette technologie. Les éléments latéraux préfabriqués sont pourvus d'un système d'éclairage intégré et une rambarde en acier inox. De la poudre de calcaire ajoutée au ciment a permis d'obtenir une couleur blanche.

From left to right, from above to below:
Supports in the river bed, detail, view from below.
Right: View of the bridge.

Von links nach rechts, von oben nach unten:
Stützen im Flussbett, Detail, Blick von unten.
Rechts: Blick auf die Brücke.

De gauche à droite, du haut vers le bas:
Vue des piles, détail d'une pile,
tablier vu d'en dessous.
À droite: Vue d'ensemble.

JERUSALEM LIGHT RAIL BRIDGE,
JERUSALEM, ISRAEL

SANTIAGO CALATRAVA ARCHITECT & ENGINEER

www.calatrava.com

Type: Cable-stayed bridge, **Length:** 360 m, **Completion:** 2008, **Photos:** Oliver Schuh+Barbara Burg / www.palladium.de, **Plans:** Santiago Calatrava.

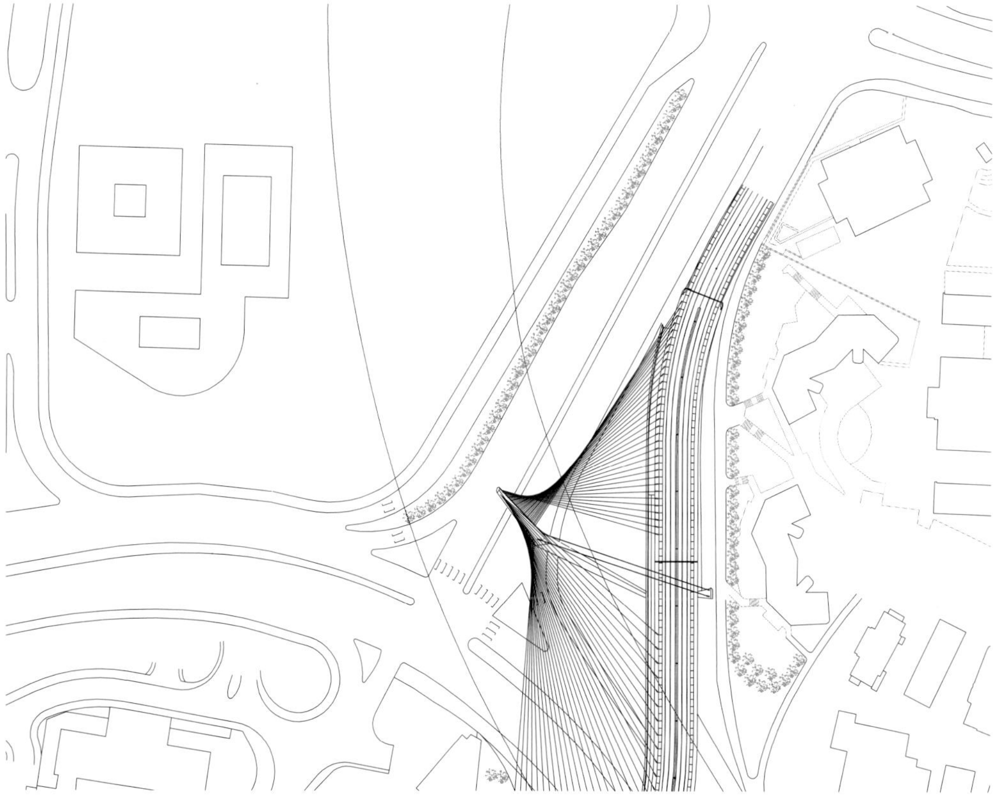

Left: Curvature. Links: Brückenschwung. À gauche: Vue de la courbe du pont. | Right: Site plan. Rechts: Lageplan. À droite: Plan de situation.

Planned in 1999 by the municipality of Jerusalem and consisting of eight lines, this bridge was designed as part of the light rail mass transit system. The light rail bridge rises over a busy traffic intersection, curving in a C-shape, as it extends from Jaffa Road to Herzl Boulevard. To accommodate this difficult location, Calatrava designed a cable-stayed bridge, with the cables anchored to a single sloping pylon sited at the center of the "C". The pylon is exceptionally light and slender, drawing attention to the cables that resemble the strings of a harp. The bridge will create a public plaza underneath it while serving as a new gateway to the city.

Diese Brücke ist Teil des acht Bahnlinien umfassenden Straßenbahnnetzes für Jerusalem, mit dessen Planung 1999 von der Stadtverwaltung begonnen wurde. Die Bahnbrücke formt ein C während sie sich über einen dicht befahrenen Verkehrsknotenpunkt erhebt und von der Jaffa Road zum Herzl Boulevard führt. Um der schwierigen Lage gerecht zu werden, gestaltete Calavatra eine Schrägseilbrücke, die nur von einem Pylon in der Mitte des C getragen wird. Die außergewöhnlich schlanke Gestalt des Pylon lenkt die Aufmerksamkeit eher auf die Seile, die in Harfenform angeordnet sind. Die Brücke bildet einen öffentlichen Platz unter ihrer Fahrbahn aus und dient gleichzeitig als neues Tor zur Stadt.

Construit au-dessus d'un carrefour routier important, ce pont est destiné à l'une des huit lignes de trains urbains prévues par la municipalité de Jérusalem depuis 1999. Sa forme en C permet de relier Jaffa Road à Herzl Boulevard. Afin de ménager l'environnement urbain, l'architecte a conçu un ouvrage suspendu à des haubans accrochés à un unique pylône incliné, positionné au milieu du C. La forme particulièrement élancée du pylône dévie l'attention vers les haubans, dont l'agencement évoque la lyre du roi David. Une esplanade va être aménagée sous le pont, le site étant appelé à devenir l'une des nouvelles portes de la ville.

On the bridge. Auf der Brücke. Ancrage des haubans.

From left to right, from above to below:
Illuminated floor panels,
Bridge curvature, pylon, pylon base.
Right: Elevation

Von links nach rechts, von oben nach unten:
Beleuchtete Bodenplatten,
Brückenschwung, Pylon, Pylonansatz.
Rechts: Ansicht.

De gauche à droite, du haut vers le bas:
Tablier illuminé, vue aérienne,
pylône et haubans, base du pylône.
À droite: Vue d'ensemble.

THREE BRIDGES FOR REGGIO NELL'EMILIA,
REGGIO EMILIA, ITALY

SANTIAGO CALATRAVA
ARCHITECT & ENGINEER

www.calatrava.com

Type: Arched bridge and cable-stayed bridges, **Length:** central bridge 221 m, lateral bridges 179 m, **Completion:** 2007, **Photos:** Oliver Schuh+Barbara Burg / www.palladium.de, **Plans:** Santiago Calatrava.

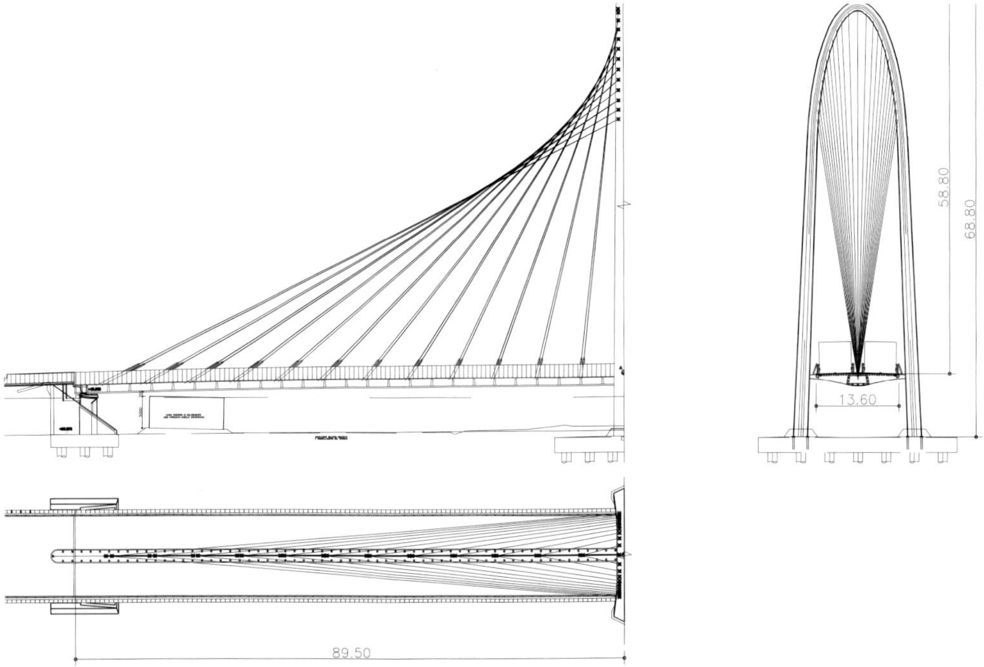

Left: Central bridge. Links: Mittlere Brücke. À gauche: Pont central. | Right: Elevation, plan and section central bridge. Rechts: Ansicht, Grundriss und Schnitt durch die mittlere Brücke. À droite: Élévation, plan et coupe longitudinale du pont central.

In October, 2007, an ensemble of three arched, cable-stayed bridges was inaugurated in Reggio nell'Emilia, as the first part of a larger infrastructural project. Plans for the bridges originated in 2002, when a new railway station was designed. In addition, a master plan for the surrounding areas of the city was created, to improve access to the site and provide an impressive new entrance from the north. This plan called for adding three new bridges to the station area, designed to connect the Autostrada del Sole (A1) to the city by a three-lane avenue. All three bridges are built of white-painted steel and reinforced concrete.

Im Oktober 2007 wurde in Reggio nell'Emilia eine Gruppe dreier bogenförmiger Schrägseilbrücken als erster Teil eines umfassenden Infrastrukturprojektes eingeweiht. Die Pläne für die Brücken erwuchsen aus dem Entwurf eines Bahnhofes im Jahre 2002, dessen umgebender Stadtteil durch einen Gesamtplan für den Straßenverkehr besser zugänglich gemacht werden sollte. Des Weiteren sollte ein imposanter Zugang von der Nordseite geschaffen werden. Dieser Plan verlangte, den Bahnhof mit drei Brücken zu verbinden und eine Anbindung der Autostrada del Sole (A1) über eine dreispurige Straße zu ermöglichen. Alle drei Brücken wurden aus weiß gestrichenem Stahl und Stahlbeton gefertigt.

Trois ponts haubanés faisant partie d'un vaste projet d'aménagement du territoire ont été inaugurés à Reggio Emilia en octobre 2007. Ils avaient été commandités en 2002, en même temps que la nouvelle gare des trains à grande vitesse, le schéma directeur stipulant que l'accès pour les véhicules automobiles arrivant du nord devait être facilité et aménagé de manière grandiose. Les trois nouveaux ponts de Reggio Emilia, entièrement réalisés en acier et béton armé peints en blanc, relient ainsi directement la ville et la gare à l'autoroute A1 (Autostrada del Sole).

From left to right, from above to below:
Central bridge, detail of stays, lateral bridge
Right: On the lateral bridge.

Von links nach rechts, von oben nach unten:
Mittlere Brücke, Detail Verspannung, seitliche Brücke.
Rechts: Auf einer der seitlichen Brücken.

De gauche à droite, du haut vers le bas:
Le pont central vu dans l'axe du tablier,
détail des haubans, un des deux ponts latéraux.
À droite: Un des deux ponts
latéraux vu au niveau du tablier.

QUARTO PONTE SUL CANAL GRANDE,
VENICE, ITALY

SANTIAGO CALATRAVA
ARCHITECT & ENGINEER

www.calatrava.com
Type: Arched bridge, **Length:** 101.2 m, **Completion:** 2008, **Photos:** Oliver Schuh+Barbara Burg / www.palladium.de, **Plans:** Santiago Calatrava.

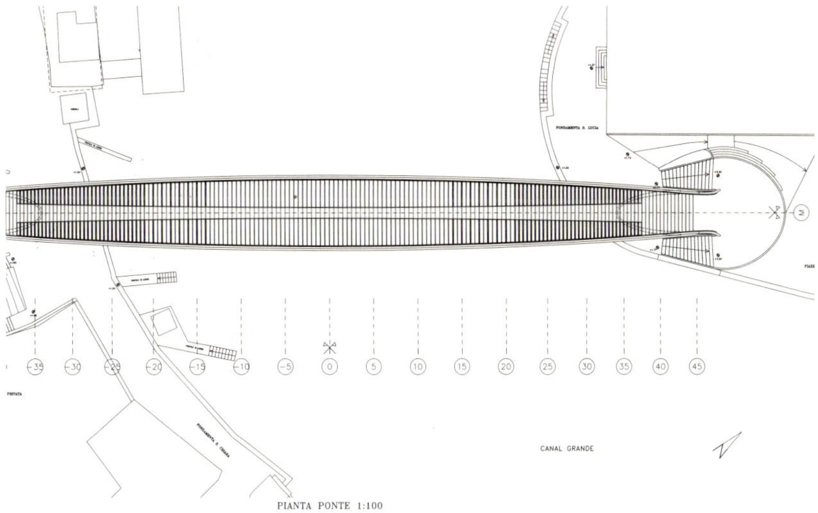

Left: Total view. Links: Gesamtansicht. À gauche: Vue d'ensemble. | Right: Plan. Rechts: Plan. À droite: Plan.

No other city in the world has as many large and small bridges as Venice. The addition of one more should therefore not be very impressive – except that the construction of a new pedestrian bridge over the Grand Canal, only the fourth to be built since the 16th century, is clearly an international event. The project is located at a strategic point, connecting the railway station (at the northern end of the bridge) to the Piazzale Roma on the south. The bridge will therefore have both functional and symbolical significance, giving visitors their first impressions of Venice and providing a panoramic view of the Grand Canal. Special care has been given to the integration of the bridge into the quays on either side.

Keine andere Stadt in der Welt verfügt über so viele Brücken wie Venedig. Eine weitere Brücke sollte demnach keinen großen Eindruck machen – es sei denn, diese neue Fußgängerbrücke über den Canale Grande ist der erst vierte Brückenneubau seit dem dem 16. Jahrhundert und wird als internationales Ereignis gefeiert. Die Brücke verbindet den Bahnhof mit der Piazzale Roma. Sie hat somit eine funktionale und symbolische Bedeutung, da sie Besuchern einen ersten Eindruck der Stadt vermittelt und einen Ausblick auf den Canale Grande gewährt. Besonders große Aufmerksamkeit wurde der Integration des filigranen und niedrigen Bogens in die bestehenden Anlegestellen auf der anderen Seite gewidmet.

Venise est la ville au monde qui compte le plus grand nombre de ponts. L'addition d'un ouvrage supplémentaire, a priori banale, ne l'est aucunement car la nouvelle passerelle sur le Grand Canal est seulement le quatrième pont construit dans la ville depuis le XVIe siècle. Elle se trouve à un endroit stratégique, reliant la gare de Venise, située au nord, à la Piazzale Roma, située au sud. La nouvelle passerelle remplit ainsi une fonction à la fois pratique et symbolique, puisqu'elle permet aux voyageurs d'accéder à la ville, tout en leur offrant d'emblée une vue panoramique sur le Grand Canal. Les architectes ont particulièrement soigné l'intégration de l'ouvrage aux quais situés de part et d'autre du canal.

Steps on the bridge. Treppenstufen auf der Brücke. Vue de la passerelle et du canal.

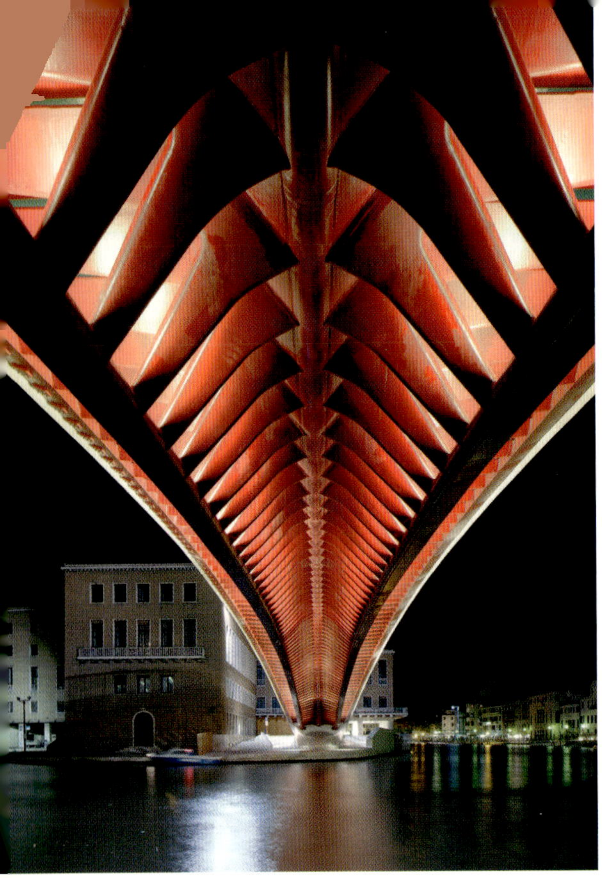

From left to right, from above to below:
View from below, bridgehead, vista.
Right: Bridge bearing.

Von links nach rechts, von oben nach unten:
Blick von unten, Brückenkopf, Ansicht.
Rechts: Auflager.

De gauche à droite, du haut vers le bas:
La passerelle vue par en dessous,
tête de pont, vue dans l'axe du canal.
À droite: Culée et passerelle.

AMANOGAWA BRIDGE,
TOKAICHI MILLENIUM FOREST, JAPAN

RINTALA EGGERTSSON ARCHITECTS

www.rintalaeggertsson.com

Type: Lounge bridge, **Length:** longest span 5 m, total legth 12.5 m, **Completion:** 2007, **Photos:** Tommaso Nova.

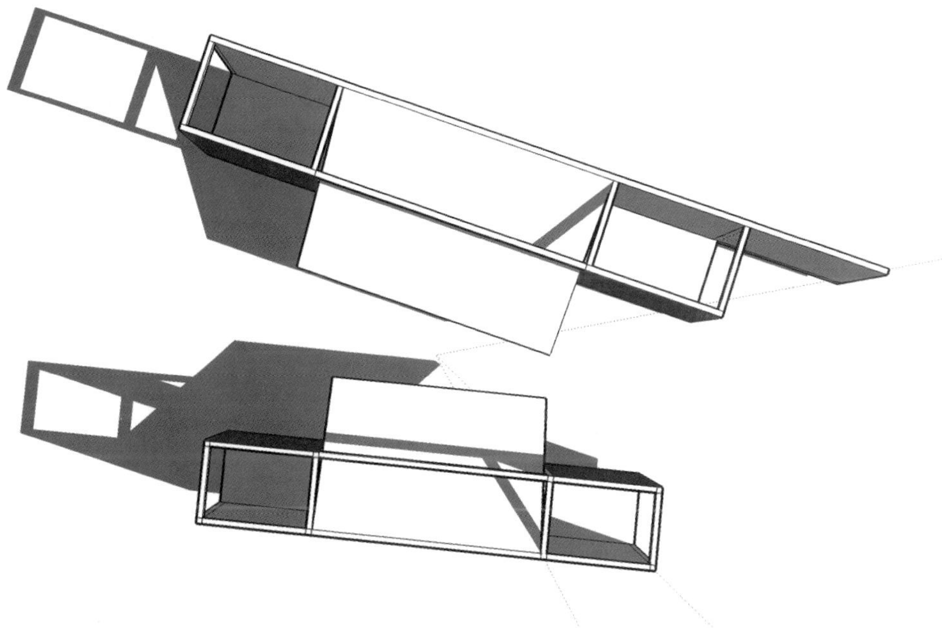

Left: Perspective through viewing slot. Links: Blick durch Sichtschlitz. À gauche: Vue au travers d'une ouverture. | Right: Roof plan. Rechts: Dachplan. À droite: Axonométrie.

The Amanogawa bridge was designed as a place of appreciation of the natural landscape surrounding it. It is a construction that enables the movement through the forest but also invites passersby to stop and to take notice of the small creek. Amanogawa is the Japanese term for the Milky Way and can literally be translated as "river of the sky." The bridge is constructed as a set of large wooden boxes that lead over the creek and provide views through several windows and viewing slots. The perception of the creek is thus altered and the attention is focused on spots that might have been ignored otherwise.

Die Amanogawa Brücke wurde als Ort der Naturbeobachtung und -bewunderung gestaltet. Sie ist eine Konstruktion, die einerseits die Bewegung durch den Wald ermöglicht, aber auch zum Verweilen einlädt, um dem kleinen Flüsschen Aufmerksamkeit zu zollen. Das japanische Wort Amanogawa bezeichnet die Milchstraße und kann wörtlich als „Fluss des Himmels" übersetzt werden. Die Brücke besteht aus großen kastenförmigen Holzgehäusen, die über das Flüsschen führen und Ausblicke durch Sichtfenster und -schlitze ermöglichen. Das Flüsschen wird dadurch ganz anders wahrgenommen und Stellen werden hervorgehoben, die andernfalls wohl ignoriert würden.

L'artiste a cherché à réaliser ici une œuvre qui permette d'apprécier et d'admirer la nature environnante. Il s'agit d'une double passerelle qui enjambe un ruisseau dans une forêt et invite les promeneurs à s'arrêter un instant. Le terme amanogawa, qui désigne la Voie Lactée en japonais, signifie littéralement « rivière dans le ciel ». La passerelle se compose de grandes « boîtes » en bois où des ouvertures focalisent l'attention du passant sur des détails qu'il ne remarquerait probablement pas sans l'aide de l'installation.

From left to right, from above to below:
Different perspectives of the two
routes across the creek.
Right: Pedestrian decks over the creek.

Von links nach rechts, von oben nach unten:
Verschiedene Ansichten der zwei
Querungen über das Flüsschen.
Rechts: Stege über dem Flüsschen.

De gauche à droite, du haut vers le bas:
Détail, vue avec arbre, extérieur,
passerelle avec ruisseau et arbre.
À droite: Vue des deux
passerelles et du ruisseau.

ALMERE CYCLE AND PEDESTRIAN BRIDGE,
ALMERE, THE NETHERLANDS

RENÉ VAN ZUUK ARCHITEKTEN B.V.

www.renevanzuuk.nl

Type: Cable-stayed bridge, **Length:** 42 m, **Structural engineers:** Van de Laar, Eindhoven, **Completion:** 2004, **Photos:** Christian Richters (182, 185), Courtesy of René van Zuuk Architekten b.v. (184).

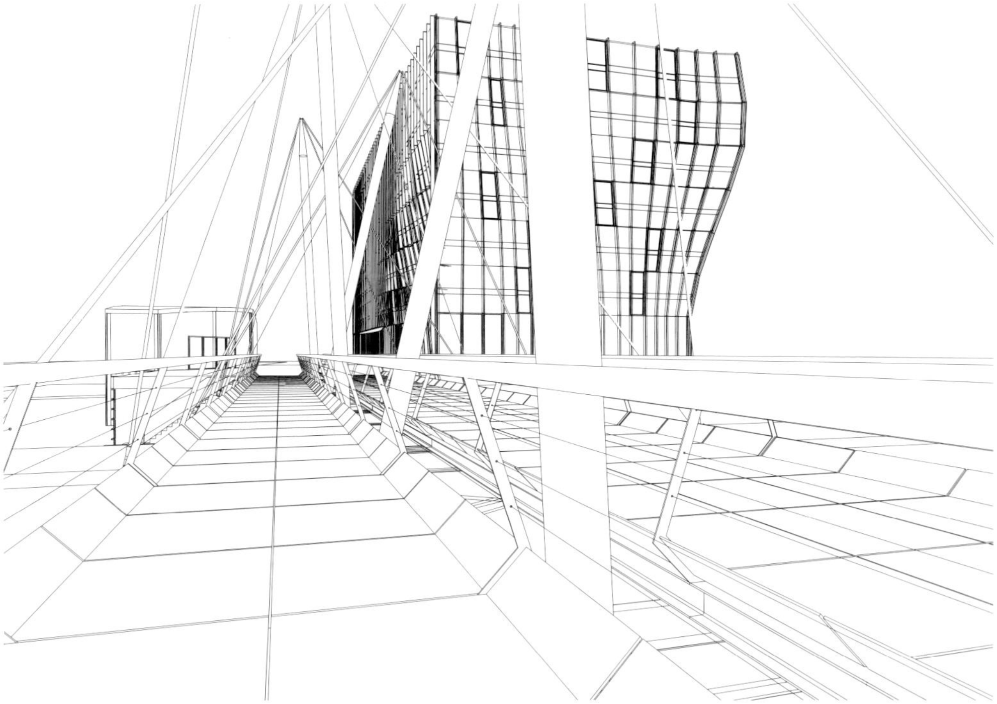

Left: View along the deck. Links: Blick entlang der Fahrbahn. À gauche: Vue d'ensemble. | Right: Perspective. Rechts: Perspektive. À droite: Vue en perspective.

The bridge spans the connecting waterway between the open Weerwater and a little harbor in the new city center of Almere. The cable-stayed mast structure refers to the capricious image of sailing boats. To enhance the slim appearance, the two traffic flows were separated. The tensegrity-like structure consists of five double beams supporting the five masts. The beams themselves are suspended from the adjacent masts with tension cables. The two fixed supports of the bridge are formed by double slanting columns on a concrete footing in the water, directly supporting the two outer beams. The eleven-meters high masts are equipped with recessed light fittings so they also serve as lighting pylons.

Die Brücke überspannt die Wasserverbindung zwischen Weerwater und einem kleinen Hafen des neuen Stadtkerns von Almere. Die Maststruktur der Schrägseilbrücke nimmt Bezug auf Segelboote. Die beiden Fahrspuren wurden unterteilt, um die schlanke Erscheinung der Brücke zu steigern. Fünf Masten getragen von fünf Doppelträgern ergeben eine selbststabilisierende Struktur. Die Träger sind mit dem jeweils benachbarten Mast durch Kabel verspannt. Die beiden Brückenstützen werden von geneigten Doppelpfeilern auf Betonfüßen im Wasser gebildet, auf denen die äußeren Träger direkt aufliegen. Die elf Meter hohen Masten sind mit eingelassenen Beleuchtungselementen versehen und dienen so auch als Lichtpylone.

Ce pont traverse un canal reliant le lac de Weerwater au petit port d'un nouveau quartier résidentiel d'Almere. Ses haubans rappellent ceux des bateaux à voile. La séparation de la chaussée en deux tabliers distincts contribue à alléger la silhouette de l'ensemble. Composée de cinq poutres doubles supportant les mats où sont accrochés les haubans, la structure semble répondre aux principes de la « tenségérité ». Chacune des deux piles se compose d'une double colonne inclinée et d'un bloc de béton émergeant de l'eau. Les mats, hauts de onze mètres, sont pourvus d'un dispositif d'éclairage.

From left to right, from above to below:
Detail of the masts, anchoring of the
cross beam, view from above.
Right: General view.

Von links nach rechts, von oben nach unten:
Detail der Masten, Verankerung der
Verstrebung, Blick von oben.
Rechts: Ansicht.

De gauche à droite, du haut vers le bas:
Détail des mats, ancrage des haubans,
le double tablier vu de haut.
À droite: Vue du pont.

BRIDGES GLANERBEEK,
ENSCHEDE, THE NETHERLANDS

NEXT ARCHITECTS

www.nextarchitects.com

Type: Cantilever and girder bridge, **Length:** 20 m, **Structural engineers:** ABT engineering and consult, **Landscape architects:** H+N+S landscape arch., **Completion:** 2007, **Photos:** Courtesy of Next Architects.

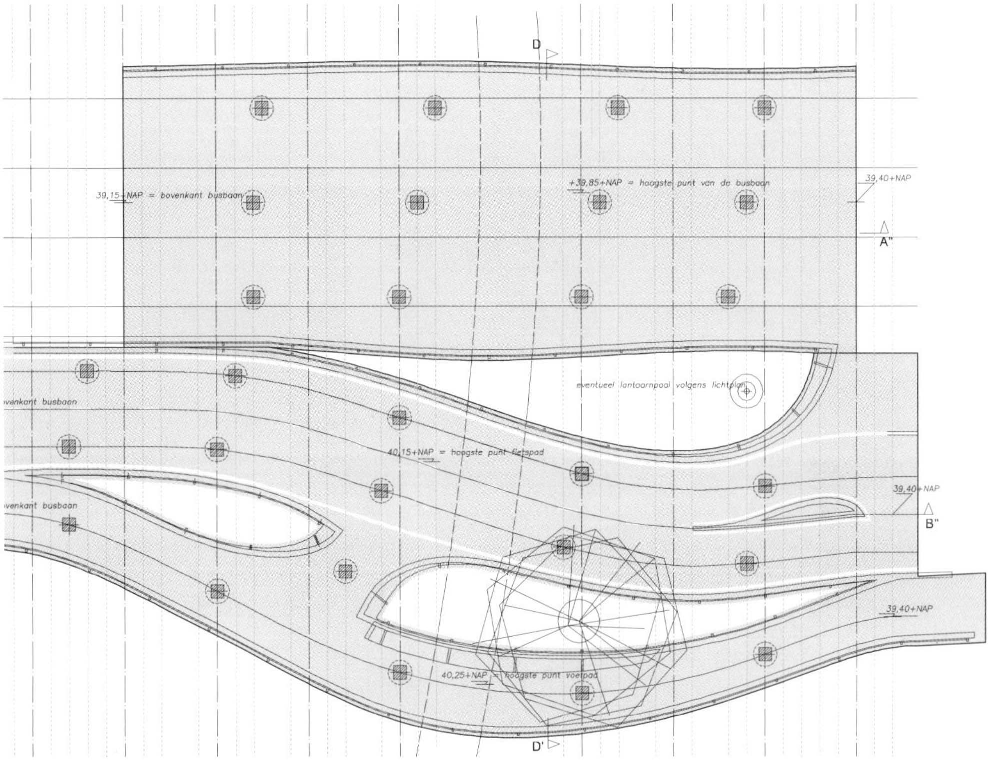

Left: Elevation. Links: Ansicht. À gauche: Vue latérale. | Right: Floor plan. Rechts: Grundriss. À droite: Plan général.

To reconcile the rather contradictory aim of transporting motor-vehicle traffic through the ecological zone of the Glanerbeek, the continuity of both was taken as a starting point. Above the brook the bridge is transformed according to visual and ecological demands. The bridge splits into three separate lanes – for buses, pedestrians and bicycles, each with its individual artery, specific maximum available height, clearance and incline. This emphasizes its transparency, providing daylight underneath the bridge for flora and fauna. A bench combines the traditional function of the bridge as a point of transition with that of a meeting place.

Um das eher widersprüchlich erscheinende Ziel einer Straßenverkehrsführung in Einklang mit der ökologischen Zone zu erreichen, wurde die Brücke über den Bach nach optischen und ökologischen Erfordernissen umgestaltet. Durch ihre Aufteilung in drei getrennte Spuren erhalten Busse, Fußgänger und Fahrräder eine jeweils eigene Verkehrsader mit spezifischer maximal zulässiger Höhe, Begrenzung und Neigung. Die so erzielte Transparenz versorgt Flora und Fauna unter der Brücke mit Tageslicht. Eine Bank erweitert die traditionelle überleitende Funktion einer Brücke zu einem Ort der Begegnung.

Les architectes devaient ici résoudre une contradiction : construire un pont automobile dans une réserve naturelle. Ils ont opté pour un style qui prend en compte aussi bien les aspects esthétiques que les critères écologiques. Le pont se compose de trois voies distinctes : la première réservée aux bus, la seconde aux piétons et la troisième aux vélos, chacune ayant une largeur et une pente qui lui sont propres. Cette conception particulière garantit d'autre part un éclairage suffisant pour la survie de la flore et de la faune aquatiques. Un banc perpétue la tradition du pont en tant qu'élément de transition et point de rencontre.

Between the lanes. Zwischen den Fahrspuren. Entre les deux voies.

From left to right, from above to below:
View from the road, pillars,
seating element on the highest point.
Right: Under the bridge.

Von links nach rechts, von oben nach unten:
Blick von der Straße, Pfeiler,
Sitzelement auf dem höchsten Punkt.
Rechts: Unter der Brücke.

De gauche à droite, du haut vers le bas:
Les ponts vus de la rue, pilier,
banc à l'endroit le plus élevé du pont.
À droite: Les ponts vus d'en dessous.

THREE BRIDGES OVER THE HOOFDVAART,
HAARLEMMERMEER, THE NETHERLANDS

SANTIAGO CALATRAVA ARCHITECT & ENGINEER

www.calatrava.com

Type: Cable-stayed bridges, **Length:** Lyre: 148.8 m, Harp, 142.83 m, Lute: 26.25 m, **Completion:** 2004, **Photos:** Oliver Schuh+Barbara Burg / www.palladium.de, **Plans:** Santiago Calatrava.

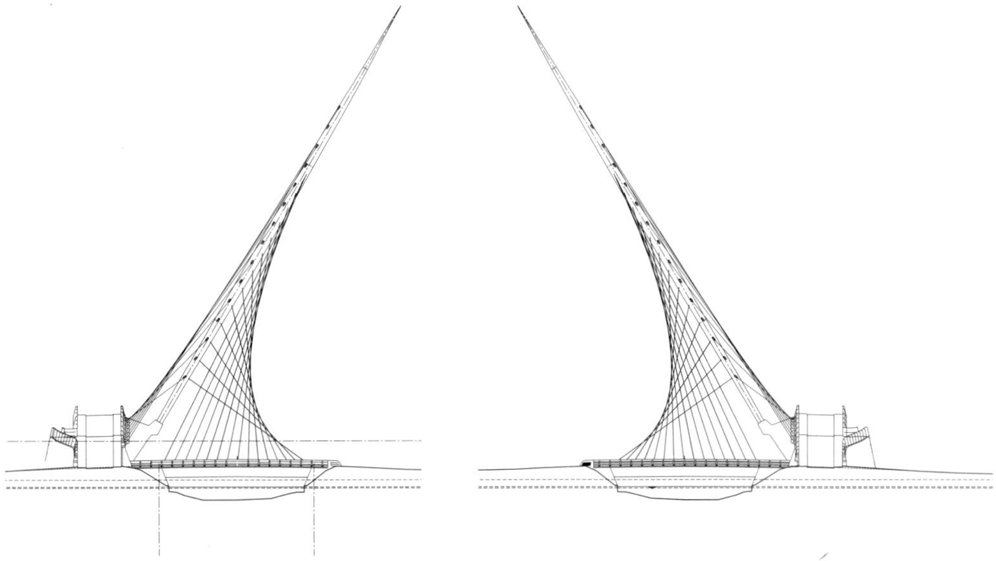

Left: Approach with pylon. Links: Brückenauffahrt mit Pylon. À gauche: Pont et pylône. | Right: Elevation of pylons. Rechts: Ansicht der Pylone. À droite: Vue des pylônes en élévation.

As part of the infrastructural enhancement of this fast economic growth area, it was decided to create bridges at main crossings over the Hoofdvaart Canal, which acts as the main drainage of what was once the Haarlem Sea. The three bridges smooth the flow of pedestrian, bicycle, and motor traffic on the local and interregional level and also serve as landmarks and as signature structures for this important new municipal area. The designs respond both to the landscape (which is very flat, with birch trees and poplars) and to the growth of the city. The trees required the addition of further vertical landmarks, with the canal itself acting as the major horizontal element.

Dieses Gebiet schnellen wirtschaftlichen Wachstums soll infrastrukturell verbessert werden. Es wurde beschlossen, dass als Teil dieser Verbesserungen Brücken an Hauptquerungen des Hoofdvaart Kanals, dem Hauptentwässerungskanal des ehemaligen Haarlemmer Meeres, errichtet werden. Die drei Brücken erleichtern den Verkehrsfluss und dienen als Wahrzeichen sowie als strukturprägende Elemente für dieses wichtige neue Viertel. Die Gestaltung rekurriert auf die flache Landschaft mit ihren Birken und Pappeln aber auch auf das Wachstum der Stadt. Die Bäume am Ufer verlangten nach weiteren vertikalen Strukturen, während der Kanal selbst als horizontales Hauptelement anzusehen ist.

Trois ponts sur le canal de Hoofdvaart, qui draine l'ancienne « mer de Haarlem », doivent améliorer l'infrastructure d'une zone urbaine en développement rapide. Ces trois ouvrages qui facilitent la circulation des piétons, vélos et véhicules à moteur sont emblématiques du quartier récemment construit à proximité. Les architectes se sont efforcés de les intégrer à l'espace urbain et au paysage uniformément plat. Dans ce sens, les pylônes sont des éléments verticaux qui viennent s'ajouter aux hêtres et aux peupliers des environs, tandis que le canal forme l'élément horizontal dominant.

From left to right, from above to below:
Different perspectives of pylon structure.
Right: View at night.

Von links nach rechts, von oben nach unten:
Verschiedene Ansichten der Pylonstruktur.
Rechts: Nachtansicht.

De gauche à droite, du haut vers le bas:
Différentes vues des pylônes.
À droite: Un pont la nuit.

NATURE BRIDGE ZANDERIJ CRAILOO,
HILVERSUM, THE NETHERLANDS

VISTA LANDSCAPE AND URBAN DESIGN

www.vista.nl

Type: Girder bridge, **Length:** 800 m, **Planning partners:** Arcadis, **Completion:** 2006, **Photos:** Goois Natuurreservaat (198 a. l., 198 a. r.), © foto W.H. Metz / Goois Natuurreservaat (196, 198 b., 199), **Plans:** Courtesy of Vista landscape and urban design.

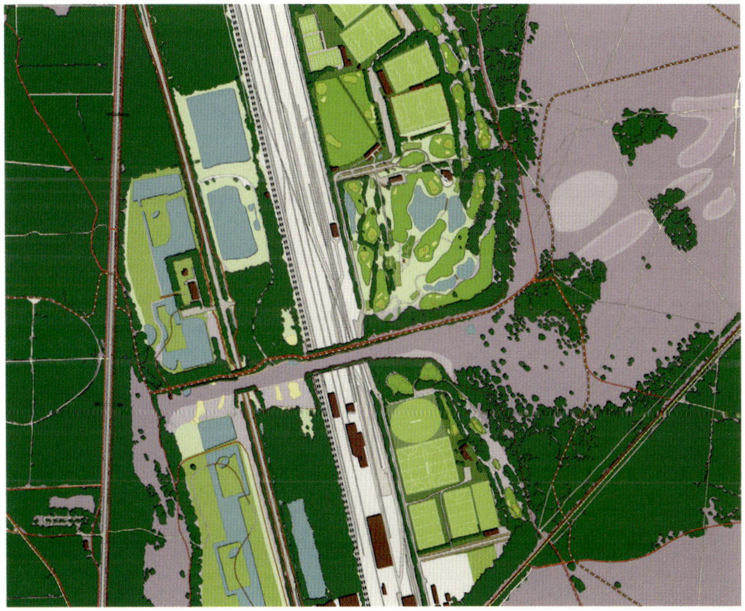

Left: Aerial view looking north. Links: Luftaufnahme Richtung Norden. À gauche: Vue aérienne en direction du nord. | Right: Master plan. Rechts: Masterplan. À droite: Schéma directeur.

A former sand quarry is the site of Europe's longest nature bridge. Connecting heather and forest landscapes over a major road and a railway yard, it provides a corridor for flora and fauna as well as walkers, cyclists and horse riders. To the west, fields were transformed into a nature education trail with a new visitors center and observation platform. Water and islands with various heights and maintenance systems result in a broad variety of habitats. To the east, existing sport facilities have been converted to accommodate the new bridge. A golf course acts as intermediary between the natural and cultural functions and forms part of the ecological corridor.

Eine ehemalige Sandgrube ist der Platz für Europas längste Naturbrücke. Dieser Korridor für Flora und Fauna sowie Radfahrer, Reiter und Fußgänger verbindet Heidelandschaften und Wald über einer Hauptstraße und einem Verschiebebahnhof. Zum Westen hin wurden Felder zu einem Naturlehrpfad mit Besucherzentrum und Aussichtsplattform umgewandelt. Wasser und Inseln in verschiedenen Höhen und mit unterschiedlichen Instandhaltungssystemen ergeben eine große natürliche Vielfalt. Zum Osten hin wurden bestehende Sporteinrichtungen umgewandelt, um sie an die neue Brücke anzupassen. Ein Golfplatz vermittelt zwischen den natürlichen und kulturellen Nutzen und bildet einen Teil des ökologischen Korridors.

Le plus long « pont naturel » d'Europe, corridor pour la faune et la flore utilisable par les piétons, les cyclistes et les cavaliers, traverse une autoroute, une voie de chemin de fer et une ancienne carrière de sable. Il assure la liaison entre deux entités distinctes : à l'ouest, d'anciennes terres agricoles reconverties en parc naturel diversifié avec plan d'eau complété par un centre d'information et une plate-forme d'observation ; à l'est, un complexe sportif réaménagé, avec notamment un terrain de golf qui sert d'interface entre la nature et les zones anthropiques.

From left to right, from above to below:
On the nature bridge, school children
on excursion, aerial view looking west.
Right: Aerial view looking north-east.

Von links nach rechts, von oben nach unten:
Auf der Naturbrücke, Schulkinder bei einem
Ausflug, Luftaufnahme Richtung Westen.
Rechts: Luftaufnahme Richtung Nordosten.

De gauche à droite, du haut vers le bas:
« Pont naturel » et autoroute en contrebas,
groupe scolaire en excursion dans le parc,
vue aérienne en direction de l'ouest.
À droite: Vue aérienne en direction du nord-est.

MELKWEGBRUG ,
PURMEREND, THE NETHERLANDS

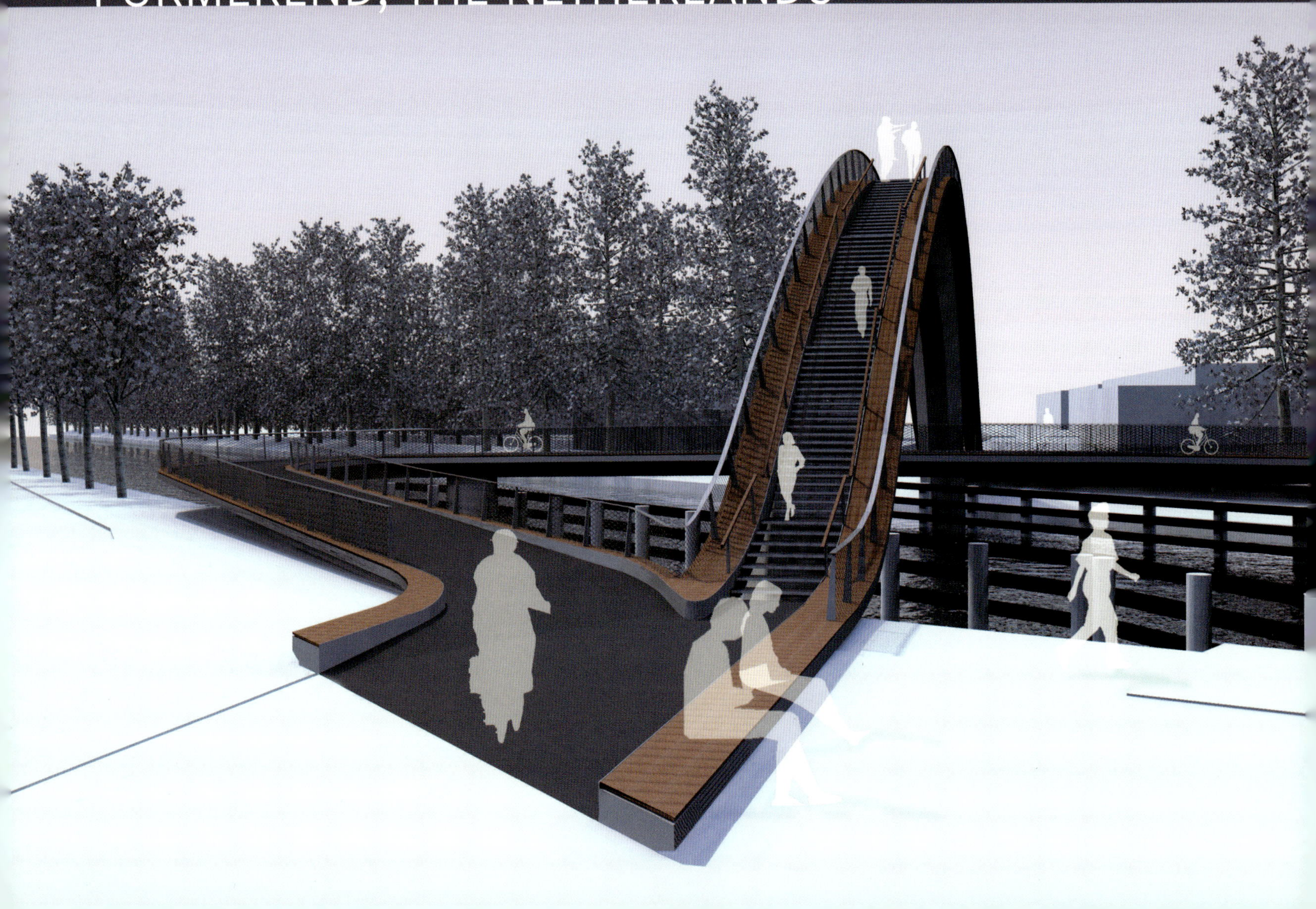

NEXT ARCHITECTS

www.nextarchitects.com

Type: Arched bridge and movable bridge, **Length:** 66 m, **Landscape Architects:** Ronald Rietveld landscape architects, **Completion:** 2011, **Photos:** Courtesy of NEXT architects.

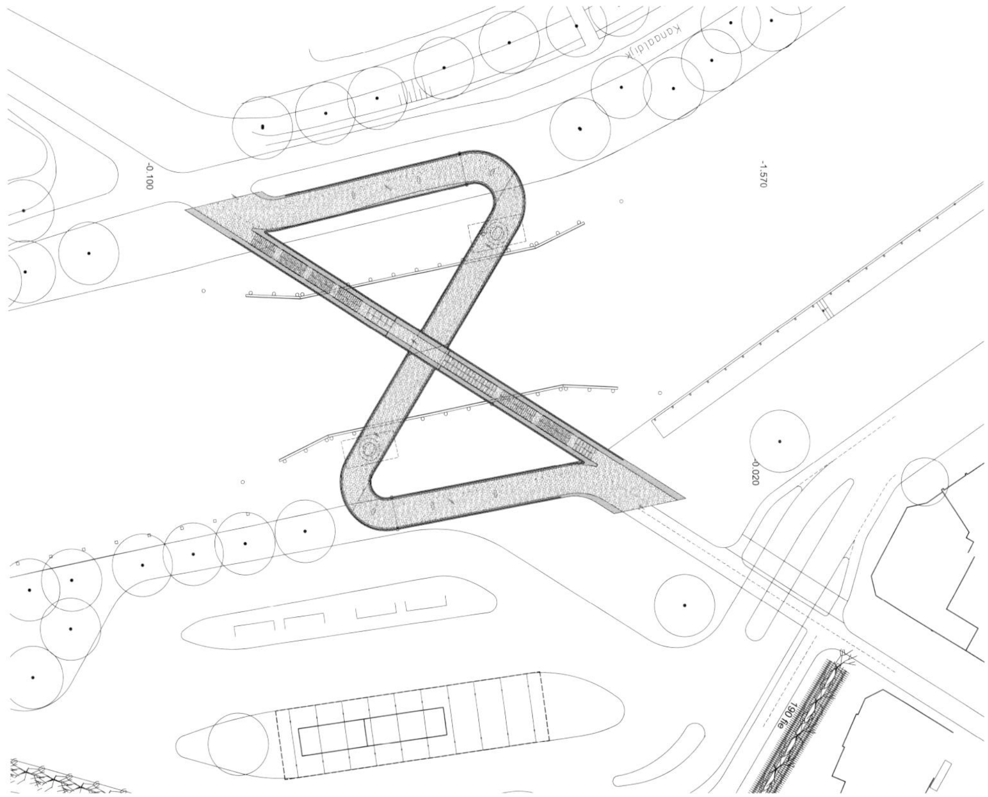

Left: View of the pedestrian arch from the square. Links: Blick vom Platz zum Fußgängerbogen. À gauche: Vue d'ensemble de la double passerelle. | Right: Site plan. Rechts: Lageplan. À droite: Plan de situation.

The Melkwegbridge connects the new Weidevenne district to the historical city. A length of 100 meters is needed to conform to the angle of inclination requirements for bicycle traffic. Thus bicycle and pedestrian routes have been separated. Pedestrians cross the canal on a high arch, reach a height of over 12 meters and can enjoy the view as the connection between old and new becomes perceivable here. The bicycle bridge coils over the water to create sufficient length to limit the angle of inclination of the bicycle path. The mobile section is divided diagonally and consists of two revolving bridge decks. The bridge is thus not only the most direct crossing but also an attraction.

Die Melkwegbrücke verbindet das neue Weidevenne-Viertel mit der Altstadt. Fußgänger und Radfahrer haben separate Querungen. Für die Querung von Radfahrern wird eine Länge von 100 Metern benötigt, um die entsprechende Neigung zu erhalten. Fußgänger überschreiten den Kanal über einen hohen Bogen, erreichen eine Höhe von 12 Metern und können die Ausblicke genießen: Alt und Neu verschmelzen hier merklich. Die Fahrradbrücke windet sich über das Wasser, um die nötige Länge und minimalen Neigungswinkel zu erreichen. Der bewegliche Teil der Brücke ist diagonal unterteilt und besteht aus zwei drehbaren Brückenplatten. Die Brücke ist damit die direkteste Querung und zugleich eine Attraktion.

Cette double passerelle assure la liaison entre le centre ville historique et le nouveau quartier résidentiel de Weidevenne. Les cyclistes franchissent le canal par une rampe à faible inclinaison qui forme un « S » d'une centaine de mètres de long, tandis que les piétons peuvent emprunter une arche qui s'élève à une hauteur de douze mètres et offre une vue panoramique sur les environs, c'est-à-dire les anciens et les nouveaux quartiers. De plus, la partie basse de l'ouvrage est mobile et s'ouvre par le milieu pour laisser passer les bateaux, la passerelle devenant ainsi une attraction au cœur de la ville.

From left to right, from above to below:
View from canal, view on the seperated decks,
view towards tghe historical city.
Right: Overview canalzone.

Von links nach rechts, von oben nach unten:
Blick vom Kanal, Blick auf die getrennten Brückenrouten,
Blick in Richtung Altstadt.
Rechts: Gesamtansicht.

De gauche à droite, du haut vers le bas:
L'ouvrage vu du canal, vue aérienne,
vue en direction du centre ville historique.
À droite: Vue aérienne de l'ouvrage et de ses environs.

RIJNHAVENBRUG,
ROTTERDAM, THE NETHERLANDS

BUREAU SLA

www.bureausla.nl

Type: Movable bridge, **Length:** 188 m, **Structural engineers:** DHV Rotterdam, DELCAN Canada, **Completion:** Ongoing, **Renderings:** Courtesy of Bureau SLA.

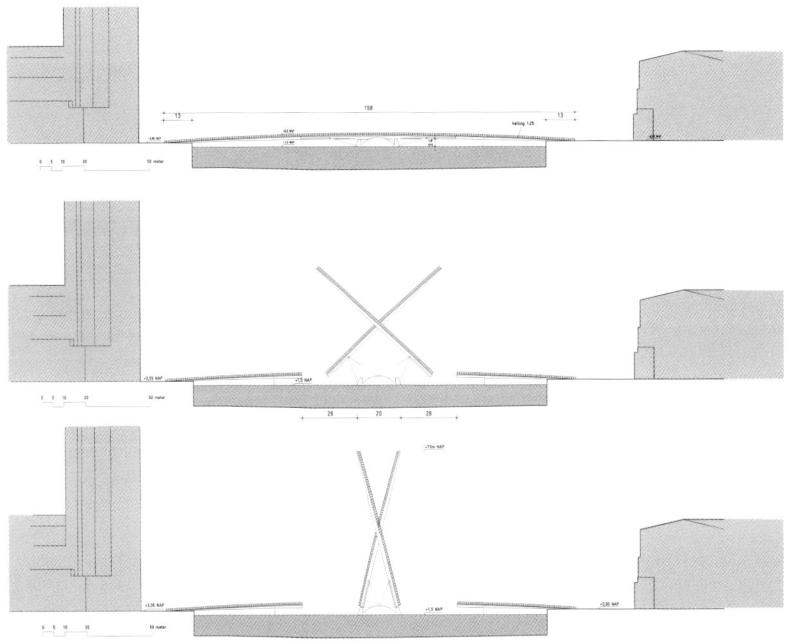

Left: View towards Wilhelminapier. Links: Blick Richtung Wilhelminapier. À gauche: Vue du pont en direction de Wilhelminapier. | Right: Elevation opening phases. Rechts: Ansicht Öffnungsphasen. À droite: Elévation (trois phases).

Wilhelminapier is a booming place. It is transforming Rotterdam into the Manhattan of the river Maas, featuring numerous designs by renowned architects. This prominent location made it necessary to give the new Rijnhavenbrug, between the Wilhelminapier and Katendrecht, a particular striking design. The architects chose a powerful and simple look for the bridge when closed, and designed a visual icon when open. Generally the bridge lies waiting quietly in the port, but a couple of times a day its particular opening movement, resembling crossed swords lifted to the sky, will capture the attention of spectators and passersby alike.

Der Wilhelminapier ist ein rasant wachsender Ort. Rotterdam wird durch ihn zum Manhattan an der Maas, wo man zahlreiche Gebäude renommierter Architekten vorfinden kann. Aus dieser Verpflichtung heraus soll die neue Rheinhafenbrücke zwischen dem Wilhelminapier und Katendrecht besonders eindrucksvoll werden. Für den geschlossenen Zustand wählten die Architekten einen kraftvollen, jedoch einfachen Ausdruck. Im geöffneten Zustand soll die Brücke aber einem Kultsymbol gleichen. So liegt sie meistens wartend im Hafen, aber einige Male am Tag wird ihre spektakuläre Öffnungsbewegung, die an zum Himmel erhobene Schwerter erinnert, alle Aufmerksamkeit auf sich ziehen.

Le quartier de Wilhelminapier, à Rotterdam, connaît un développement remarquable et est en passe de devenir le « Manhattan des bouches de la Meuse », avec des immeubles neufs réalisés par les plus grands architectes du moment. Dans ce contexte, le pont qui relie Wilhelminapier au quartier de Katendrecht se devait d'être exceptionnel. Le concept retenu se caractérise par sa simplicité lorsque le pont est en position normale, et son apparence spectaculaire lorsqu'il se relève pour laisser passer un navire plusieurs fois par jour. Gageons que la partie centrale du tablier, se relevant en deux parties qui se croisent dans le ciel comme des épées géantes, ne manquera pas d'attirer l'attention du public.

From left to right, from above to below:
Opening phase 1 (resting), opening phase 2,
opening phase 3, view from bridge deck.
Right: Opening of the bridge.

Von links nach rechts, von oben nach unten:
Öffnungsphase 1 (geschlossen), Öffnungsphase 2,
Öffnungsphase 3, Blick vom Brückendeck.
Rechts: Öffnen der Brücke.

De gauche à droite, du haut vers le bas:
position fermée (phase 1), début de l'ouverture (phase 2),
position ouverte (phase 3), passants sur le pont.
À droite: Ouverture du pont.

RANDSTADRAIL STATION BEATRIXKWARTIER,

THE HAGUE, THE NETHERLANDS

ZWARTS & JANSMA ARCHITECTS

www.zwarts.jansma.nl

Type: Lattice bridge, **Length:** 400 m, **Planning partners:** Gemeentewerken Rotterdam, **Completion:** 2006, **Photos:** DigiDaan, Amsterdam.

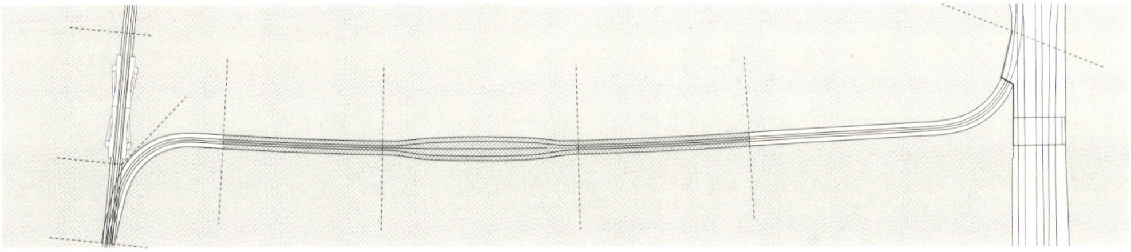

Left: General view. Links: Gesamtansicht. À gauche: Vue d'ensemble. | Right: Plan of the station. Rechts: Plan des Bahnhofs. À droite: Plan de la gare.

Randstadrail is the project of a new light urban rail network for the area between The Hague and Rotterdam. The space-frame tubular construction of the viaduct follows the curve that urban planner Joan Busquets has set out for the alignments and the roadways on the Beatrixlaan. At a length of 400 meters, the viaduct is constructed from a skeleton structure of rings of constructional steel strips with a diameter of about ten meters, interconnected by diagonally set tubes to form an open tube structure providing room for two train tracks. Thanks to the big spans between the V-shaped columns there is hardly any visual obstruction at eye level, so that social and traffic safety are not compromised.

Als Teil von Randstadrail, einem Projekt, welches das Schienensystem Den Haags und Rotterdams verbinden soll, folgt die Raumfachwerkröhre des Viadukts der Kurve, die Stadtplaner Joan Busquets für Ausrichtung und Straßenführung der Beatrixlaan vorgesehen hat. Auf einer Länge von 400 Metern besteht es aus einer Skelettstruktur aus zehn Meter durchmessenden Ringen von Baustahlstreifen, die mittels diagonal verlaufender Röhren verbunden werden. So entsteht ein weitgehend geöffnetes Rohr, das zwei Gleise aufnehmen kann. Die V-förmigen Stützen stehen weit auseinander, so dass sich ebenerdig kaum Störungen des Blicks, des Verkehrs und der sozialen Interaktion ergeben.

Un nouveau chemin de fer urbain à structure légère relie La Haye à Rotterdam. La portion présentée ici est un viaduc tubulaire qui suit le tracé de l'avenue Beatrixlaan déterminé par l'urbaniste Joan Busquets. Long de quatre cents mètres et d'une largeur suffisante pour des trains roulant dans les deux sens, l'ouvrage se compose de cercles en acier de dix mètres de diamètre reliés entre eux par des tubes positionnés en diagonale. Ce type de construction permet d'espacer les supports en forme de « V » de manière à réaliser de grandes portées et à ne pas obstruer la vue ni gêner la circulation au niveau du sol.

From left to right, from above to below:
Detail columns, track from below,
framework, nodal points.
Right: View into the station.

Von links nach rechts, von oben nach unten:
Detail Stützen, Trasse von unten,
Fachwerkumhüllung, Knotenpunkte.
Rechts: Bahntrasse mit Blick in den Bahnhof.

De gauche à droite, du haut vers le bas:
Supports, viaduc vu d'en dessous,
vue en perspective, points nodaux.
À droite: Vue d'une voie et de la gare.

BRIDGE VLAARDINGSE VAART,
VLAARDINGEN, THE NETHERLANDS

WEST 8 URBAN DESIGN & LANDSCAPE ARCHITECTURE B.V.

www.west8.com

Type: Lattice bridge, **Length:** 45 m, **Structural engineers:** ABT / Rob Nijsse, **Completion:** 2009, **Photos:** Courtesy of West 8.

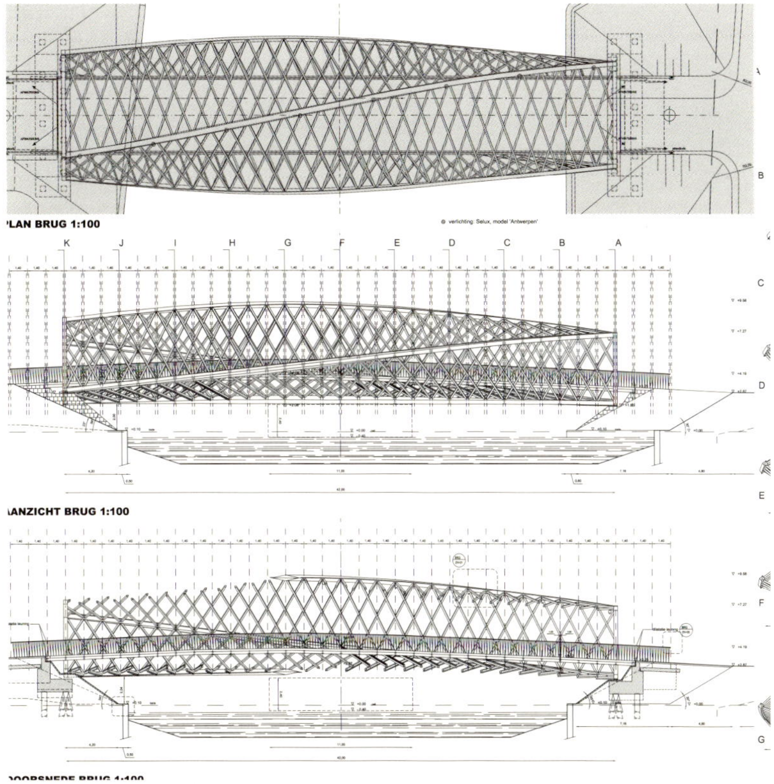

Left: View from the river bank. Links: Blick vom Ufer. À gauche: Vue d'ensemble. | Right: Plan, section and view. Rechts: Grundriss, Schnitt und Ansicht. À droite: Plan, élévation et vue en coupe.

The bridge in Vlaardingen is a crucial connection within the urban and recreational bicycle network. The expressive sculpture establishes the moment of 'crossing' and by so doing captures the qualities of the immediate landscape. Its construction comprises 400 red steel tubes that were welded together to create a unique and dynamic structure. The three-dimensional truss of the bridge twists on its horizontal axis. This twisted space frame construction absorbs the user in a continually changing experience of perspective. The rectangular frame at the bridge entrance evolves into a double-height diamond form in the middle. A transparent cathedral of steel becomes a new resting and meeting place.

Die Brücke in Vlaardingen bildet eine entscheidende Verbindung im Radwegenetz der Stadt und des Erholungsgebietes. Ihre skulpturale Form verweist auf den Moment des Überquehrens und greift dabei die Vorzüge der Landschaft auf. Die Konstruktion besteht aus 400 rot gestrichenen, verschweißten Stahlrohren, die eine einzigartige dynamische Struktur hervorbringen. Das dreidimensionale Pfeilerwerk windet sich in seiner horizontalen Achse und zieht so die Nutzer in ein Erlebnis ständig variierender Perspektiven. Der rechteckige Rahmen am Brückenzugang entwickelt sich zur Mitte hin zu einer Kristallform doppelter Höhe. Eine transparente Kathedrale aus Stahl wird zum neuen Rastplatz und Treffpunkt.

Ce pont joue un rôle important dans le réseau de pistes cyclables de la ville de Vlaardingen et de la région alentour. Sa forme torsadée en fait une œuvre paysagère de qualité sculpturale. L'ouvrage, d'apparence particulièrement dynamique, se compose de quatre cents tubes d'acier soudés entre eux et peints en rouge. La torsion axiale crée pour les usagers du pont des perspectives toujours changeantes. Les cadres rectangulaires positionnés horizontalement à chacune des extrémités doublent de taille au milieu du pont, où ils forment un losange posé sur la pointe. Cette « cathédrale transparente » en tubes d'acier a d'emblée conquis le cœur des promeneurs.

From left to right, from above to below:
Railing, inside the bridge, general view.
Right: Detail.

Von links nach rechts, von oben nach unten:
Geländer, in der Brücke, Gesamtansicht.
Rechts: Detail.

De gauche à droite, du haut vers le bas:
Gros plan sur la rambarde, fanfare
sur le pont, vue dans l'axe de la rivière.
À droite: Détail.

DIOMEDE ARCHIPELAGO,
BERING STRAIT, RUSSIA, USA

TALLER 301

www.taller301.com

Type: Cantilever and girder bridge, **Length:** 85 km, **Architects:** Pablo Forero Quintero, Julián Restrepo Molina, Manuela Mosquera Iragorri, Susana Somoza Parada, Tomás Jaramillo Valencia, **Structural engineers:** Nicolas Parra García, **Completion:** Ongoing, **Photos:** Taller 301 / David Dodman (216, 218 a. l.), Taller 301 / Bogotá (218 a. r.), Taller 301 / Kentish Plover (218 b.), Taller 301 / Aerometric (219).

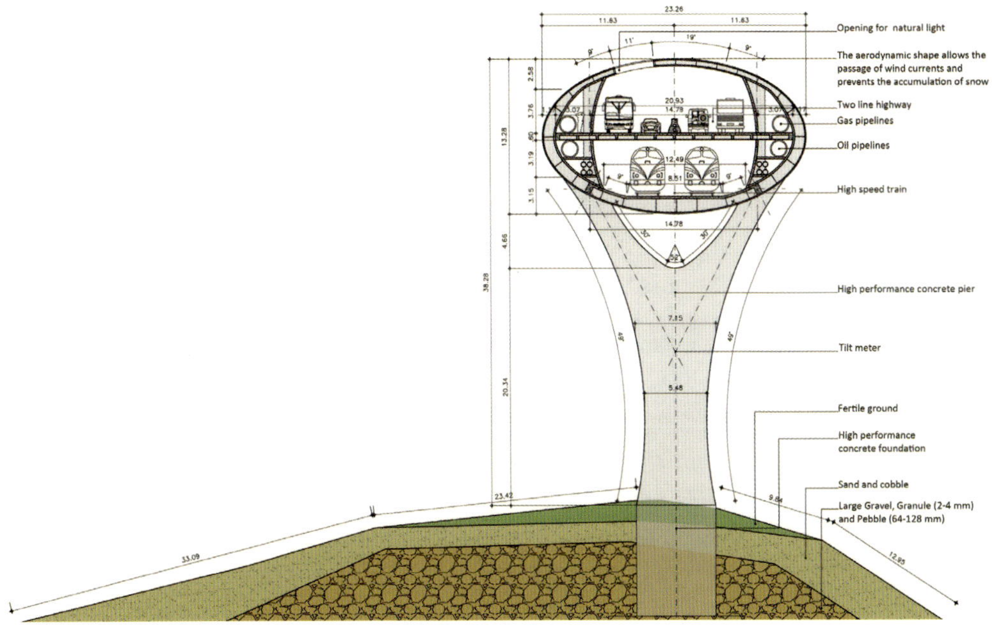

Left: Peace memorial. Links: Friedensdenkmal. À gauche: Mémorial de la Paix. | Right: Typical section. Rechts: Typischer Querschnitt. À droite: Vue en coupe.

The archipelago is composed of two intertwined land paths that meet in the Diomede Islands. One path, the infrastructure path, allows the traffic of vehicles, high speed trains and pipelines. The other path provides people interested in nature, access to the natural park. Through dredging and land reclamation processes ground was created to protect the infrastructure path from the extreme climate and weather. These land paths would be separated within the next decades by the harsh weather conditions transforming and eroding them into a group of small islands. What was artificially created would then be transformed by nature.

Das Archipel setzt sich aus zwei ineinander greifenden Landzungen zusammen, die sich bei der Diomedes-Insel treffen. Einer der Landwege, der Infrastruktur-Weg, ermöglicht Individualverkehr, Hochgeschwindigkeitszüge und Fernleitungen. Der andere Landweg dient Naturinteressierten als Zugang zum Park. Durch Ausgrabungen und Landgewinnungsarbeiten wird der Infrastruktur-Weg durch die erschaffenen Landmassen vor den extremen Witterungseinflüssen geschützt. Ohne diesen Schutz würden die Landzungen innerhalb der nächsten Jahrzehnte durch Umwelteinflüsse getrennt, da durch Erosion die Wege in kleinere Inseln zerfallen würden. Somit würde das künstlich Geschaffene von natürlichen Bedingungen zerstört.

Un projet baptisé « Archipel de Diomède » vise à construire un pont au travers du détroit de Bering. L'ouvrage, qui portera un pipeline et sera accessible aux véhicules automobiles et aux trains à grande vitesse, se doublera d'une structure destinée aux personnes voulant découvrir le parc naturel aménagé des deux côtés du détroit. Des travaux de dragage et de remblai seront nécessaires pour assurer la solidité de l'ensemble dans cet environnement soumis à des conditions climatiques extrêmes. À défaut de quoi l'érosion naturelle détruirait en quelques décennies l'ouvrage titanesque réalisé de main d'homme.

From left to right, from above to below:
Bridge in the archipelago, bridge
during arctic melting, on the coast.
Right: Aerial view.

Von links nach rechts, von oben nach unten:
Brücke im Archipel, Brücke zur Zeit
der Eisschmelze, an der Küste.
Rechts: Luftaufnahme.

De gauche à droite, du haut vers le bas:
Pont sur l'archipel artificiel, détroit
lors de la fonte des glaces, la côte.
À droite: Vue aérienne.

PUENTE DE L'ASSUT D'OR,
VALENCIA, SPAIN

SANTIAGO CALATRAVA
ARCHITECT & ENGINEER

www.calatrava.com

Type: Cable-stayed bridge, **Length:** 180 m, **Completion:** 2008, **Client:** C.A.C., S.A. / Generalitat Valenciana, **Photos:** Oliver Schuh + Barbara Burg / www.palladium.de, **Plans:** Santiago Calatrava.

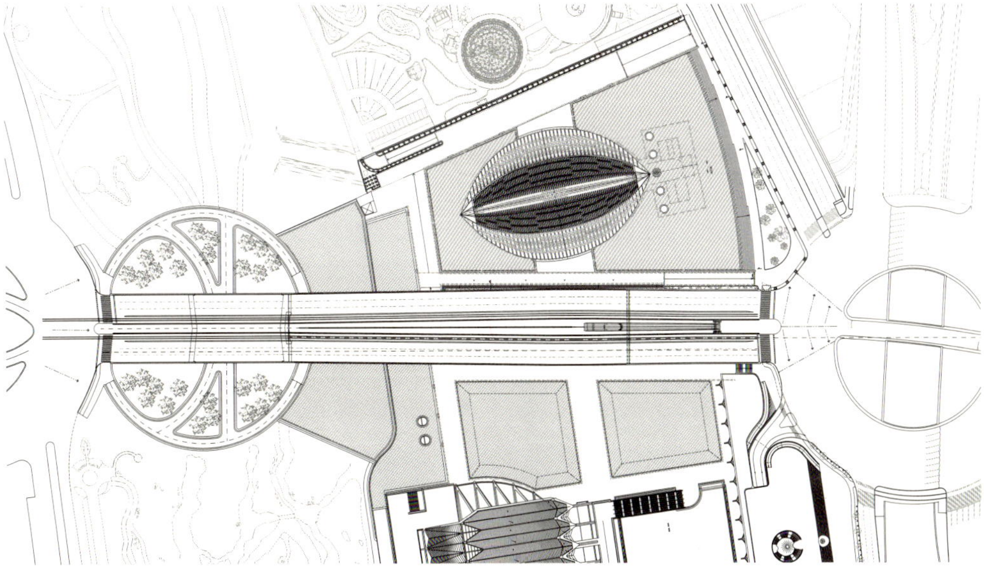

Left: Bridge with its distinctive pylon. Links: Brücke, mit ihren auffälligen Pylonen. À gauche: Le pont et son pylône caractéristique. | Right: Site plan. Rechts: Lageplan. À droite: Plan de situation.

Puente de L'Assut d'Or (the former Serrería Bridge) will stand along the easternmost edge of Santiago Calatrava's City of Arts and Sciences, within the dry bed of the Turia River. Integrated into the design of the City of Arts, the new bridge serves as a landmark for that site, while also providing a much-needed crossing for vehicles and pedestrians. Its structure is that of a cable-stayed bridge, whose deck is suspended from an inclined curved pylon built on the southern part of the site. The deck spans 180 meters and varies in total width from 35.5 meters to 39.2 meters. The pylon rises to a height of 123 meters.

Puente de L'Assut d'Or (ehemals Serreria Brücke) wird am östlichsten Rand von Santiago Calatrava's City of Arts and Sciences im trockenen Flussbett des Turia errichtet. Die neue Brücke wird durch ihre Einbindung in das gestalterische Grundmuster zum Wahrzeichen der Region, schafft aber auch dringend benötigte Verkehrswege für Fahrzeuge und Fußgänger. Ihre Struktur ist die einer Schrägseilbrücke, deren Fahrbahnplatte an einem betonummantelten, geschwungenen Pylon an der Südseite aufgehängt ist. Die Fahrbahnplatte überspannt eine Distanz von 180 Metern und variiert in der Gesamtbreite zwischen 35,5 Metern und 39,2 Metern. Der Pylon erhebt sich auf eine Höhe von 123 Metern.

Cet ouvrage (appelé également « Puente de Serrería ») borde à l'est la Cité des arts et des sciences construite par Santiago Calatrava dans le lit asséché de la rivière Turia. Ouvrage emblématique et d'un style adapté à celui de la Cité, le nouveau pont établit une liaison dont piétons, bus, voitures et camions avaient cruellement besoin. Le tablier, long de 180 mètres pour une largeur allant de trente-cinq à trente-neuf mètres, est retenu par des haubans fixés à un unique pylône incurvé de 123 mètres de haut, qui se dresse au sud de l'ouvrage.

From left to right, from above to below:
Detail of cables, worm's-eye view of pylon,
view from deck towards city core.
Right: View at night.

Von links nach rechts, von oben nach unten:
Detail der Kabel, Pylon Ansicht von unten,
Blick über das Deck Richtung Stadtkern.
Rechts: Nachtansicht.

De gauche à droite, du haut vers le bas:
Gros plan sur les haubans et le pylône,
haubans et pylône vus d'en dessous, vue dans
l'axe du pont avec la ville à l'arrière plan.
À droite: Le pont au crépuscule.

ZARAGOZA PAVILION,
ZARAGOZA, SPAIN

ZAHA HADID ARCHITECTS

www.zaha-hadid.com

Type: Girder bridge with pavilion, **Length:** 270 m, **Design:** Zaha Hadid with Patrik Schumacher, **Structural engineers:** ARUP Associates, **Completion:** 2008, **Photos:** Fernando Guerra.

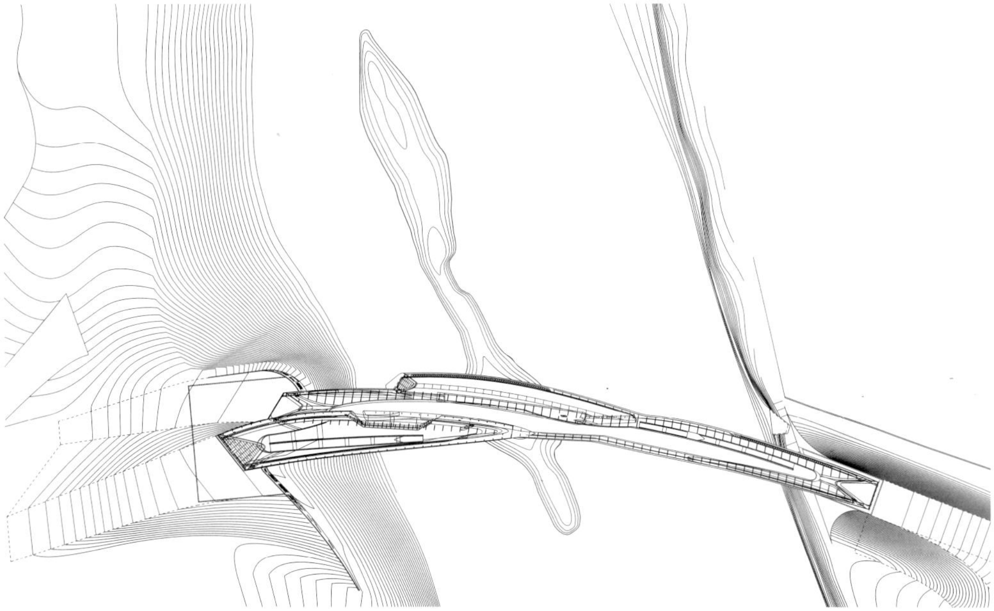

Left: Bridge entrance. Links: Brückeneingang. À gauche: Entrée. | Right: Lower floor site plan. Rechts: Lageplan der unteren Etage. À droite: Plan de situation.

Zaragoza pavilion was designed in the course of the Expo 2008 to house the exposition "Water, a unique resource" while serving as one of the main entrances. The bridge pavilion provides a pedestrian crossing across the Ebro River linking the Station Boulevard to the area of the Conference Center. Inside the dynamic-shaped structure different zones can be found each with its own spatial identity and function. Besides its breathtaking shape, the bridge pavilion comprises some striking technical features like the shark-scale skin, consisting of a permeable membrane allowing an internal microclimate based on natural airflow. The foundations of the bridge extend to a depth of 68.5 meters.

Der Pavillon wurde für die Expo 2008 als Ausstellungsraum für „Water, a unique resource" und als Zugang zum Gelände entworfen. Der Brückenpavillon bietet eine Querung über den Fluss Ebro und verbindet den Station Boulevard mit dem Konferenzbereich. Innerhalb der dynamischen Struktur sind verschiedene Zonen angelegt, mit jeweils eigener räumlicher Anordnung und Funktion. Der Brückenpavillon weist neben seiner erstaunlichen Form auch einige technische Besonderheiten auf. So zum Beispiel die durchlässige, an Haie erinnernde Außenmembran, welche das interne Raumklima auf Basis natürlicher Luftzirkulation reguliert. Die Fundamente dieser Brücke reichen bis auf eine Tiefe von 68,5 Metern.

Ce « pont pavillon » réalisé en 2008 servait d'entrée à l'exposition universelle de Saragosse et abritait une exposition consacrée à l'eau. Il enjambe l'Èbre et permet d'aller directement du boulevard de la Gare au palais des Congrès. L'intérieur de cet édifice aux formes volontaires s'organise autour d'espaces à l'atmosphère personnalisée remplissant des fonctions distinctes. Parmi les avancées technologiques intégrées à l'ouvrage, citons notamment l'enveloppe « peau de requin » composée d'une membrane perméable à l'air qui assure un microclimat à l'intérieur. Les fondations s'enfoncent jusqu'à 68,5 mètres dans le lit du fleuve.

From left to right, from above to below:
Inside the bridge, detail skin, inner structure.
Right: Bridge pavilion at night.

Von links nach rechts, von oben nach unten:
Im Inneren der Brücke, Detail der Umhüllung,
innere Struktrur der Brücke.
Rechts: Brückenpavillon bei Nacht.

De gauche à droite, du haut vers le bas:
Intérieur, détail de l'enveloppe,
autre vue de l'intérieur.
À droite: Pavillon au crépuscule.

LIMMAT FOOTBRIDGE WITH PROMENADE ELEVATOR,

BADEN, SWITZERLAND

LEUPPI & SCHAFROTH ARCHITEKTEN

www.leuppischafroth.ch

Type: Lattice bridge, **Length:** Bridge 52 m, walkway 16 m, tower 35 m, **Structural engineers:** Henauer Gugler AG, **Artist:** Beat Zoderer, **Completion:** 2007, **Photos:** Roger Frei, Zurich.

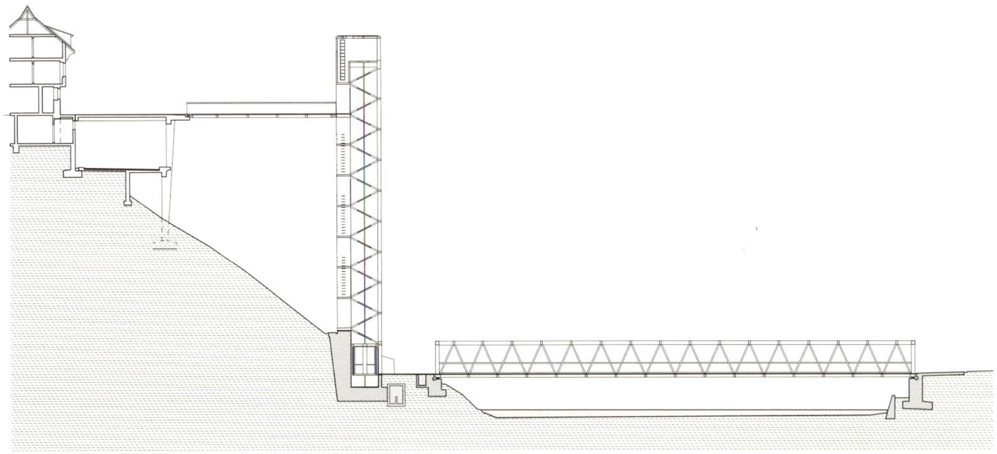

Left: View from south-east, Ennetbaden. Links: Blick von Südost, Ennetbaden. À gauche: Passerelle inférieure vue du sud-est (Ennetbaden). | Right: Section. Rechts: Schnitt. À droite: Vue en coupe.

The Limmat River meanders around the city of Baden, forming a plateau that naturally separates it from the neighboring town of Ennetbaden. This geological feature impedes the connection between the two towns. As such, a direct access for pedestrians and bicyclists has been needed for decades. An accessible steel structure was erected on the location of the former cable ferry. This new linear path is composed of a horizontal bridge, a vertical elevator tower and a horizontal walkway. Red-brown in color – varying in shade depending on the light – the steel artifact complements its surroundings. The ensemble shines at night with indirect illumination.

In Baden windet sich die Limmat um das Stadtplateau und formt einen naturnahen Flussraum. Die Verbindungen von Ennetbaden nach Baden sind durch diesen geologischen Einschnitt auf eine natürliche Weise erschwert. Nach einer neuen Verbindung für Fußgänger und Radfahrer wurde schon lang gesucht. Am Ort der früheren Seilfähre wurde mit einer geradlinigen Wegefolge eine begehbare Plastik aus Stahl errichtet; gegliedert in einen liegenden Brückenkörper, in einen stehenden Aufzugsturm und in eine liegende Passerelle. Rotbraun – wechselnd temperiert vom Tageslicht – wirkt die Materialität des Stahles naturnah und archaisch. Nachts strahlt sie - inszeniert durch eine indirekte Beleuchtung.

La rivière Limmat forme un méandre autour du plateau où se trouve la ville de Baden. Cette configuration géographique rendait difficile la liaison entre le village d'Ennetbaden et la localité voisine. C'est pourquoi une passerelle est venue récemment remplacer le bac qui existait depuis longtemps. Il s'agit d'une structure linaire en acier qui franchit la rivière et se complète d'un ascenseur et d'une seconde passerelle. L'ensemble s'intègre parfaitement à l'environnement grâce à sa couleur rouge-brun quelque peu archaïque. Un éclairage indirect met l'ouvrage en valeur à la nuit tombée.

View from south. Ansicht von Süden. L'ouvrage vu du sud.

From left to right, from above to below:
Limmatpromenade viewed from south, on the bridge,
view from north-east, general view.
Right: Walkway.

Von links nach rechts, von oben nach unten:
Limmatpromenade von Süden, auf der Brücke,
Blick von Nordost, Gesamtanlage.
Rechts: Passerelle.

De gauche à droite, du haut vers le bas:
Promenade le long de la rivière vue du sud,
vue dans l'axe, la passerelle
vue du nord-est, vue d'ensemble.
À droite: Passerelle supérieure.

MAILLART BRIDGE RESTORATION,
WINTERTHUR, SWITZERLAND

DR. DEURING + OEHNINGER AG
ENGINEERS ETH SIA USIC

www.deuringoehninger.ch

Type: Arched bridge, **Length:** 48 m, **Planning partners:** Stutz + Bolt + Partner Architekten AG, **Original:** 1933 by Robert Maillart, W. Pfeiffer and A.-G. Baugeschäft Wülflingen, **Completion:** 2000, **Photos:** Dr. Martin Deuring.

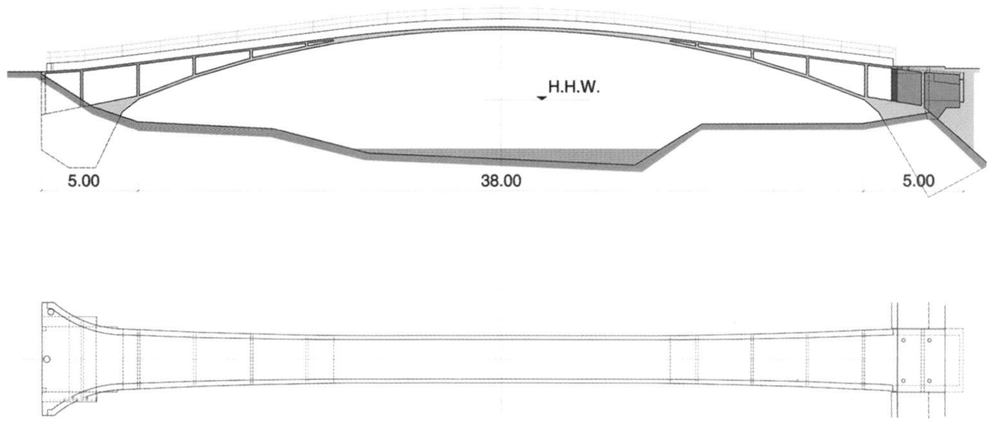

Left: General view. Links: Gesamtansicht. À gauche: Vue d'ensemble. | Right: Elevation and plan. Rechts: Ansicht und Grundriss. À droite: Élévation et plan.

First built in the year 1933, this 48-meter long bridge with art history significance was restored in the year 2000. An important premise for the work was maintaining the structure's shape, location and materials. Following extensive appraisals of the existing structure, the necessary measures were determined. They included improvement of the static situation by adding a concrete underbed to the abutment near the adjacent street, the removal of two layers of mortar, the exposing of the railway posts to a depth of 40 millimeters, and the application of reinforcement protection as well as crack-covering protective covers.

Im Jahr 2000 wurde diese 48 Meter lange und bereits seit 1933 bestehende, kunsthistorisch bedeutende Brücke in Stand gesetzt. Wichtiges Kriterium bei den Arbeiten war die Erhaltung des Bauwerkes in seiner Form, Lage und Materialwahl. Nach gründlichen Untersuchungen am Bestand wurden die notwendigen Maßnahmen festgelegt. Diese umfassten die Verbesserung der statischen Situation durch Unterbetonieren des Widerlagers im Bereich der angrenzenden Straße, den Abtrag zweier bestehender Mörtelschichten, das Freilegen der Geländerpfosten auf eine Tiefe von 40 Millimetern und das Aufbringen eines Armierungsschutzes sowie rissüberdeckender Schutzschichten.

Cette passerelle de quarante-huit mètres de long était d'une conception révolutionnaire lorsqu'elle a été construite en 1933. Lors de sa rénovation en l'an 2000, le cahier des charges imposait de conserver la forme, la position et les matériaux d'origine, de sorte que les architectes ont dû étudier l'ouvrage en détail afin de déterminer les mesures à prendre. Celles-ci ont consisté à améliorer la statique en doublant le tablier d'une couche de béton sur la face inférieure, à supprimer deux couches de mortier, à dégager les supports de la rambarde sur une profondeur de quatre centimètres, à protéger l'armature et à combler les fissures.

From left to right, from above to below:
Bridge approach from below,
crown, construction.
Right: Perspective.

Von links nach rechts, von oben nach unten:
Brückenanlauf von unten,
Scheitelpunkt, Konstruktion.
Rechts: Perspektive.

De gauche à droite, du haut vers le bas:
Le tablier vu d'en dessous, la passerelle
vue dans l'axe, vue latérale.
À droite: Vue en perspective.

LOCHNAGAR STREET BRIDGE,
BOW, EAST LONDON, UNITED KINGDOM

KNIGHT ARCHITECTS WITH ALAN BAXTER & ASSOCIATES

www.knightarchitects.co.uk, www.alanbaxter.co.uk

Type: Concrete shell-form beam, **Length:** 55 m, **Structural engineers:** Alan Baxter & Associates, **Completion:** 2010, **Renderings:** Knight Architects Limited.

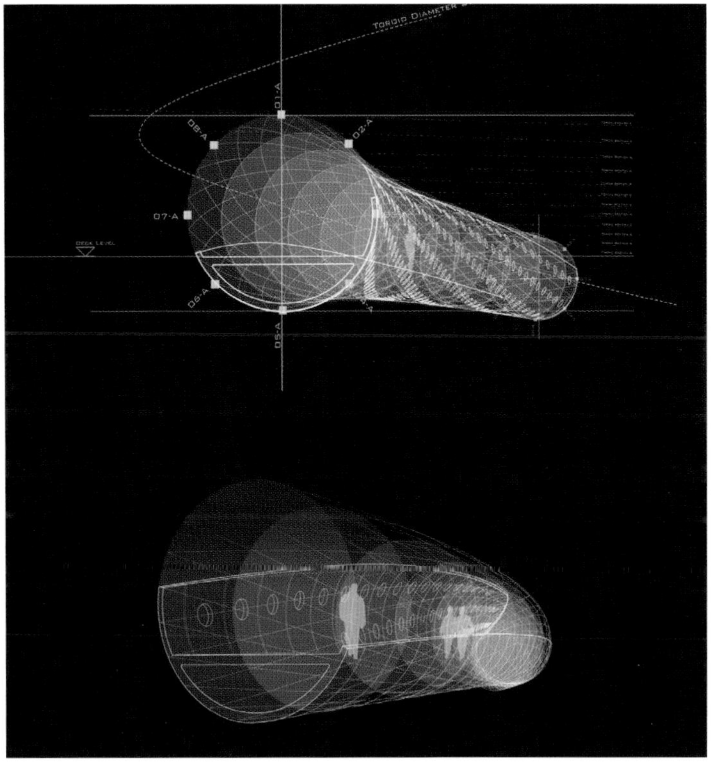

Left: View of bridge from north. Links: Ansicht der Brücke von Norde. À gauche: Passerelle vue du nord. | Right: Geometric study. Rechts: Geometrische Studie. À droite: Étude de géométrie toroïdale.

The striking shell-form bridge will provide Poplar Riverside with a dramatic new pedestrian and cycle link across the River Lee. The design provides both a functional crossing and leisure use. Both aspects are intended to help break down east-west physical barriers and facilitate residents reaching work in other city districts. The bridge is a beam whose distinctive shell form is derived from simple circular geometry, allowing it to be pre-cast using identical elements. Ultra high performance concrete permits a very thin, lightweight structure with openings for views and lighting. The bridge is designed to be pre-assembled.

Die augenfällige muschelförmige Brücke wird das Flussufer Poplars um eine neue Fußgänger- und Fahrradverbindung über den Fluss Lee erweitern. Die Gestaltung sieht eine funktionale Querung vor, die Naherholungsnutzen mit berücksichtigt. Beide Aspekte dienen der Überwindung von Ost-West-Barrieren und sollen den Anwohnern Arbeitswege in andere Stadtteile erleichtern. Die Brücke besteht aus einem Träger, dessen Muschelform von einfacher ringförmiger Geometrie abgeleitet ist, die eine Vorfertigung einzelner sich wiederholender Elemente zulässt. Sehr stabiler Beton ermöglichte die filigrane Struktur mit Sicht- und Beleuchtungsöffnungen. Die Brücke wurde so gestaltet, dass sie vormontiert werden kann.

Une passerelle pour piétons et cyclistes d'apparence peu conventionnelle traverse la rivière Lee, dans l'est de Londres. Cet ouvrage en forme de coque contribue à unifier les quartiers est et ouest de la ville en supprimant les barrières physiques et en rapprochant les riverains de leurs nouveaux lieux de travail. L'ensemble se compose d'éléments préfabriqués dérivés de la géométrie toroïdale simple, réalisés en béton hautes performances et percés d'ouvertures de manière à minimiser le poids et maximiser à la fois l'éclairage et les vues sur les alentours. Les éléments ont été préassemblés, puis amenés par une barge sur le site avant d'être mis en place définitivement.

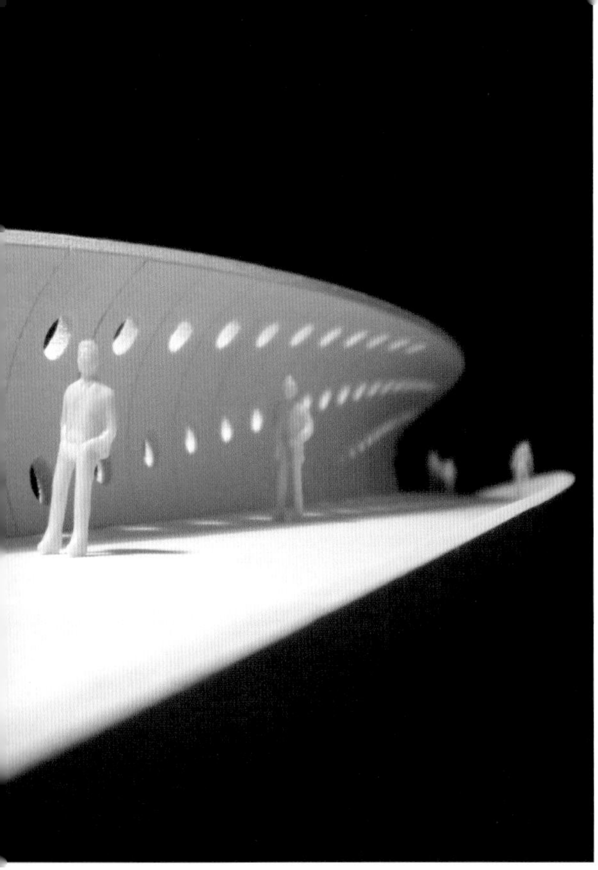
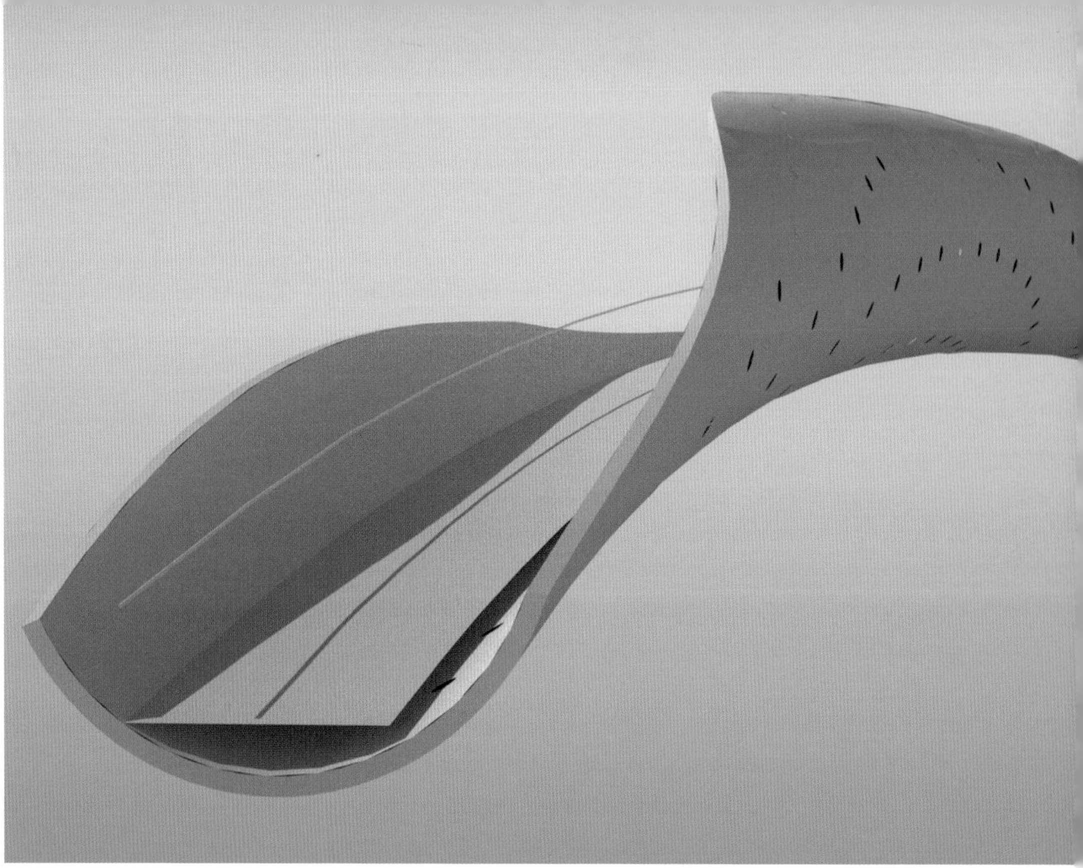
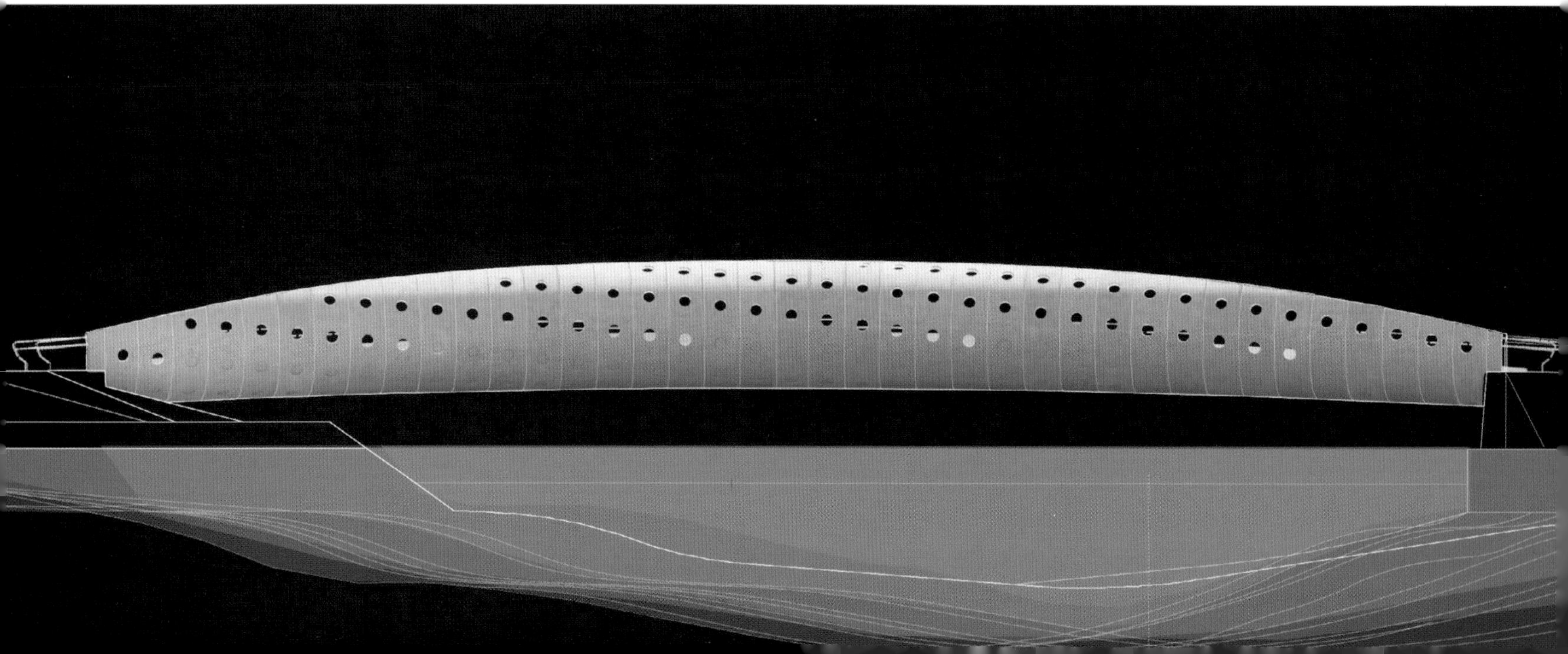

From left to right, from above to below:
Model, study, elevation.
Right: View of bridge from south.

Von links nach rechts, von oben nach unten:
Modell, Studie, Ansicht.
Rechts: Blick auf die Brücke von Süden.

De gauche à droite, du haut vers le bas:
Maquette, étude, élévation.
À droite: Passerelle vue du sud.

CASTLEFORD BRIDGE,
CASTLEFORD, UNITED KINGDOM

MCDOWELL+BENEDETTI

www.mcdowellbenedetti.com

Type: Cantilever and girder bridge, **Length:** 131 m, **Structural engineers:** Alan Baxter & Associates, **Hydrological engineers:** Arup Water, **Lighting consultant:** Sutton Vane & Associates, **Quantity surveyors:** Philip Pank Partnership, **Client:** Wakefield Metropolitan District Council, **Completion:** 2008, **Photos:** Timothy Soar.

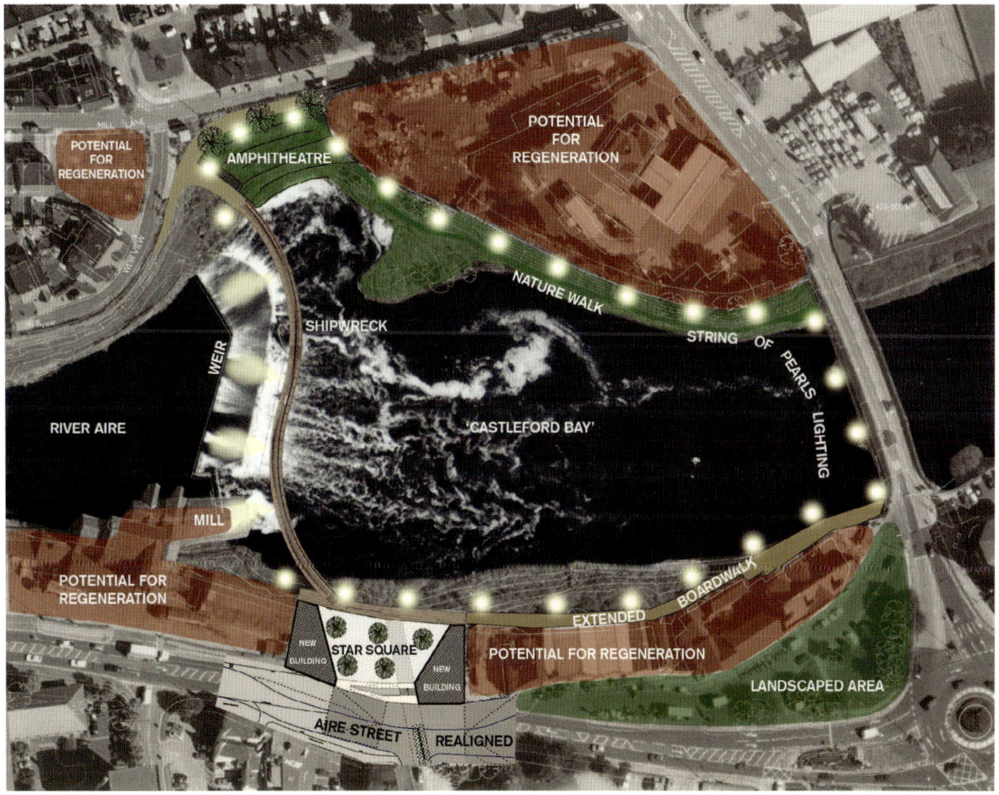

Left: Bottom side facing north. Links: Unterseite Richtung Norden. À gauche: La passerelle vue d'en dessous en direction du nord. | Right: Master plan. Rechts: Bebauungsplan. À droite: Schéma directeur.

This bridge is part of a waterfront regeneration master plan for Castleford Bay. It is not designed as a mere route, but as a generous, 131-meter long and four meter wide public space. One of its most striking features is the innovative spanning structure that rises from the deck to create four 20-meter long curving benches that invite passersby to sit down. Timber was chosen as the main material to provide a natural character and visual warmth, helping the bridge's elegant and sinuous style fit harmonically into the historic river landscape, while providing maximum visibility and a floating sensation.

Die Brücke ist Teil eines Erneuerungsplans für das Ufergebiet der Castleford Bay. Sie wurde nicht als einfacher Weg konzipiert, sondern als großzügiger öffentlicher Platz mit 131 Metern Länge und vier Metern Breite. Eines der augenfälligsten Merkmale ist die innovative Verspannung, welche sich von der Oberflächenplatte abhebt und 20 Meter lange, gewellte Bänke formt, die zum Verweilen einladen. Holz bestimmt als eines der Hauptmaterialien den natürlichen Charakter des Bauwerkes, verleiht ihm Wärme und trägt dazu bei, dass die elegante gewundene Brücke sich harmonisch in die historische Uferlandschaft einfügt, während sie ein Maximum an Ausblicken bietet und ein Gefühl des Dahinschwebens erzeugt.

Cette passerelle fait partie du schéma directeur visant à la revitalisation de la baie de Castleford. Longue de 131 mètres et large de quatre, elle a été conçue comme un élément de paysage urbain et non comme une simple voie de communication. L'ouvrage intègre diverses innovations, notamment une surélévation partielle dans l'axe du tablier de manière à former quatre bancs de vingt mètres de long d'où l'on peut admirer le paysage. Les architectes ont accordé la priorité au bois, matériau naturel qui donne à l'ensemble un aspect chaleureux. La forme élégante et sinueuse de la passerelle minimise l'impact sur le paysage, tout en lui donnant un aspect caractéristique qui n'est pas sans rappeler un tapis volant.

ALLINSONS
STONEGROUND
FLOUR

Bottom side facing south towards mill. Unterseite Richtung Mühle auf der Südseite. La passerelle vue d'en dessous en direction du sud et de la minoterie.

From left to right, from above to below:
Detail of sustainable Cumaru hardwood deck,
Cumaru detail, deck facing north, pylon.
Right: Bottom side facing south.

Von links nach rechts, von oben nach unten:
Detail des Holzdecks aus Cumaru,
Cumaru, Deck Richtung Norden, Pfeiler.
Rechts: Unterseite Richtung Süden.

De gauche à droite, du haut vers le bas:
Deux vues de détail du tablier en
bois exotique (cumaru), vue du tablier en
direction du nord, pylônes.
À droite: La passerelle vue d'en dessous
en direction du sud.

GATESHEAD MILLENNIUM BRIDGE,
GATESHEAD, UNITED KINGDOM

GIFFORD WITH WILKINSON EYRE ARCHITECTS

www.gifford.uk.com, www.wilkinsoneyre.com

Type: Arched, cable-stayed and movable bridge, **Length:** 105 m, **Completion:** 2004, **Photos:** Courtesy of Gifford.

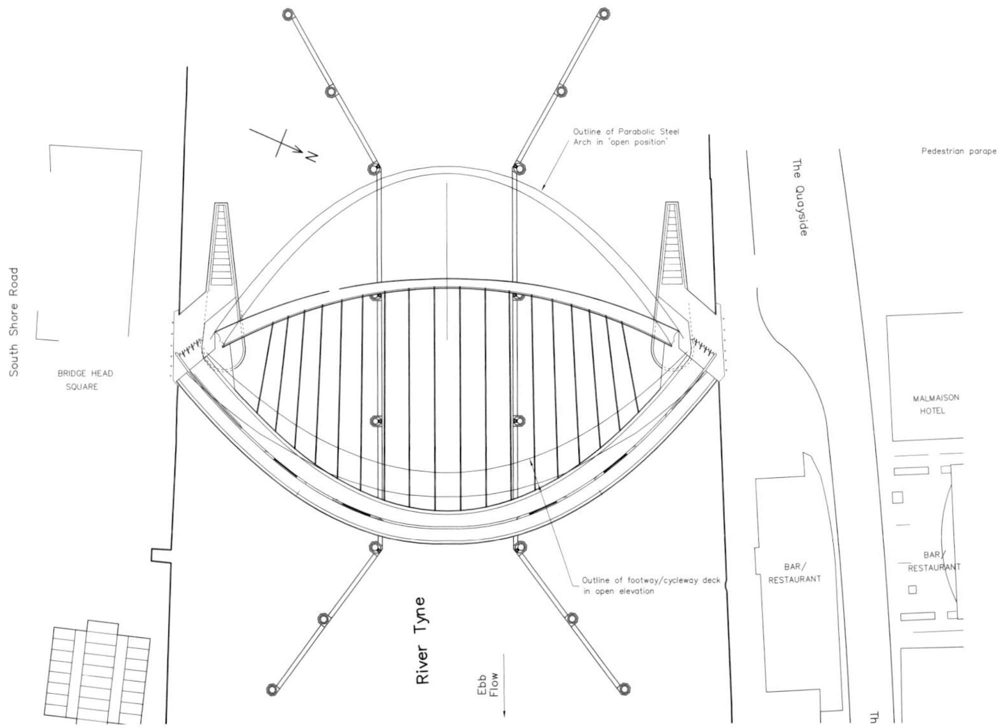

Left: Night view from the south bank. Links: Nachtansicht vom Südufer. À gauche: Le pont vu de nuit à partir de la rive sud. | Right: Floor plan. Rechts: Grundriss. À droite: Plan.

This impressive bridge was designed by Gifford in collaboration with Wilkinson Eyre Architects. Providing a new crossing for pedestrians and cyclists over the River Tyne just downstream of the existing landmark structures, it gave Gateshead a new iconic structure for the new millennium. The bridge shape, a pair of steel arches spanning 105 meters between concrete islands, can be rotated up to 40 degrees to permit vessels to pass beneath. Gateshead Millenium Bridge is England's first bridge capable of rotating around its longitudinal axis.

Diese imposante Brücke wurde von Gifford in Zusammenarbeit mit Wilkinson Eyre Architects entwickelt. Sie bildet einen neuen Übergang für Fußgänger und Radfahrer über den Fluß Tyne. Ein Stück flußabwärts der bestehenden Wahrzeichen der Stadt gelegen, ist sie als ikonische Struktur Gatesheads für das neue Jahrtausend gedacht. Die Brückenkonstruktion bestehend aus zwei Stahlbögen, überspannt 105 Meter zwischen Betoninseln, und kann bis zu 40 Grad gedreht werden, damit Schiffe passieren können. Die Gateshead Millenim Brücke war die erste Brücke Englands, die sich um ihre Längsachse drehen lässt.

Cette passerelle pour piétons et cyclistes est un ouvrage mobile particulièrement impressionnant jeté sur la rivière Tyne dans le centre ville de Newcastle au tout début du nouveau millénaire. Il consiste en deux arches d'acier d'une portée de 105 mètres qui reposent sur des piles en béton et peuvent effectuer une rotation de quarante degrés pour laisser passer les gros bateaux. Cet ouvrage d'art est ainsi le premier pont mobile construit en Angleterre avec un axe de rotation longitudinal.

BALTIC FLOUR MI
SEE IT
BE IT

Aerial view. Luftaufnahme. Vue aérienne.

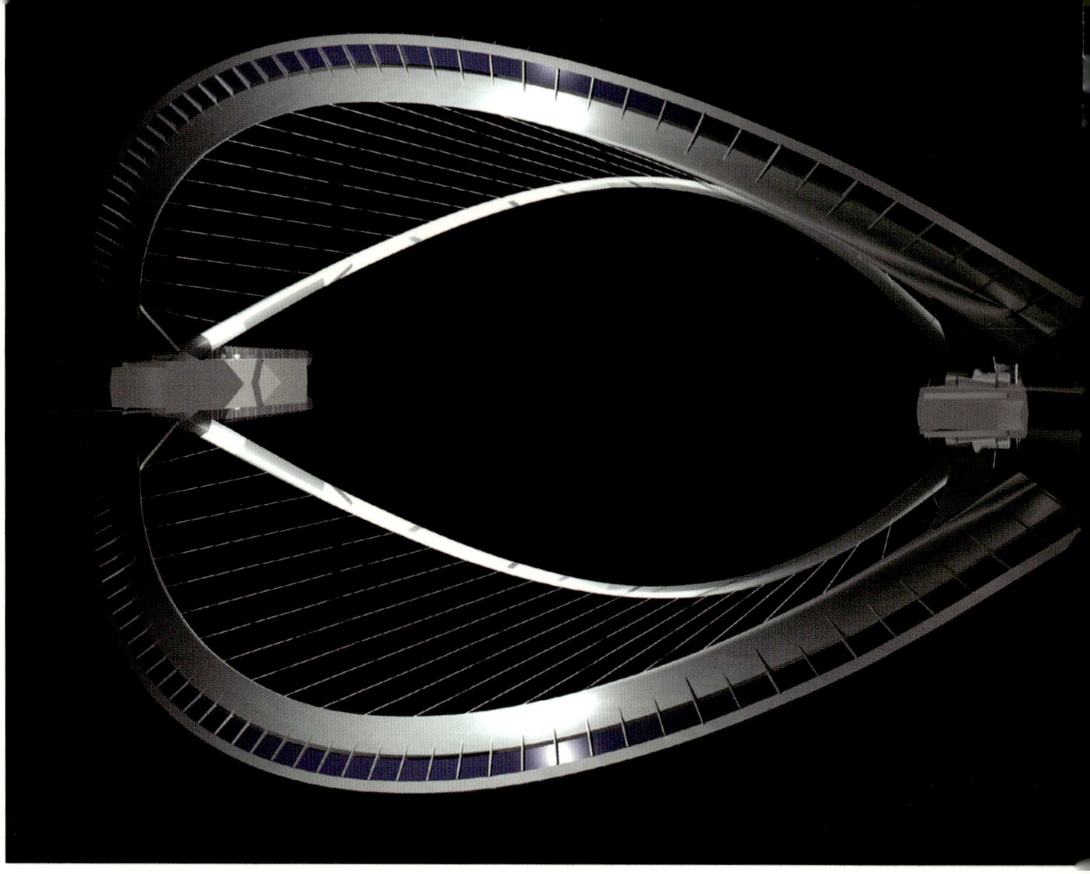

From left to right, from above to below:
View from north bank, aerial view of fully opened bridge, bridge fully open, illuminated arch.
Right: Gateshead Millennium Bridge and the Tyne Bridge beyond.

Von links nach rechts, von oben nach unten:
Blick vom Nordufer, Luftaufnahme der komplett geöffneten Brücke, geöffnete Brücke, beleuchteter Bogen.
Rechts: Gateshead Millennium Bridge vor der Tyne Bridge.

De gauche à droite, du haut vers le bas:
Pont vu de la rive nord, pont en position ouverte (dessin et vue de nuit), arche illuminée.
À droite: Vue de la rivière Tyne avec le Gateshead Millennium Bridge et le Tyne Bridge à l'arrière plan.

KENT MESSENGER MILLENNIUM BRIDGE,
MAIDSTONE, UNITED KINGDOM

STUDIO BEDNARSKI LTD.

www.studio-bednarski.com

Type: Stress ribbon bridge, **Length:** 101.5 m, **Structural engineers:** Strasky Husty and Partners, **Completion:** 2001, **Photos:** CB Bednarski, London.

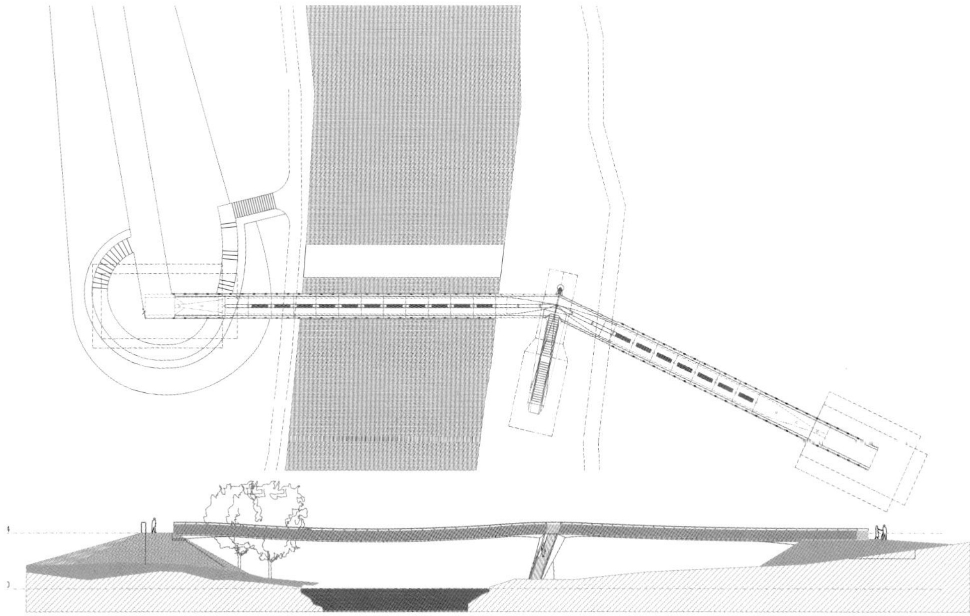

Left: Side view. Links: Seitenansicht. À gauche: Vue latérale. | Right: Plan and elevation. Rechts: Plan und Ansicht. À droite: Plan et vue en élévation.

This pedestrian bridge connects the Maidstone Millennium River Park with the town across the Medway River. It is the world's first "cranked" stress ribbon bridge and among the few stress ribbon bridges in the world. Its deck consists of two concrete catenaries designed with a dogleg alignment with a 25 degree angle between them. It appears to be the only concrete UK Millennium Bridge and the only one that was not the subject of a design competition. Among all the Millennium Bridges in the United Kingdom it was built with the lowest price per square meter of deck, including additional costs related to extreme flood conditions.

Diese Fußgängerbrücke verbindet den Maidstone Millenium River Park über den Medway River mit der Innenstadt. Sie ist die weltweit erste gekröpfte Spannbandbrücke und gehört zu den wenigen dieser Bauart in der Welt. Die Fahrbahn besteht aus zwei Betonführungen, die in gegenläufiger Ausrichtung und mit 25 Grad Winkel dazwischen gestaltet sind. Sie ist wahrscheinlich die einzige Milleniumbrücke Großbritanniens und Nordirlands aus Beton und die einzige, die ohne Wettbewerbsausschreibung beauftragt wurde. Innerhalb der Milleniumbrücken Großbritanniens und Nordirlands hat sie die geringsten Baukosten pro Quadratmeter Fahrbahn verursacht, einschließlich aller Kosten für Hochwasserschutz.

Cette passerelle au-dessus de la rivière Medway assure la liaison entre le parc du Millénaire et le centre ville de Maidstone. C'est un des rares ponts caténaires au monde et le premier présentant une structure angulaire. Le tablier se compose de deux éléments en béton formant entre eux un angle de vingt-cinq degrés. Par ailleurs, c'est le seul pont « du millénaire » britannique construit en béton et sans avoir fait l'objet d'un concours d'architecture. C'est aussi le moins cher au mètre carré de tablier, même en tenant compte des coûts additionnels liés aux risques d'inondation.

From left to right, from above to below:
Mesh parapet, view of stair prop,
rear view, view of stairs.
Right: Cranked stress ribbon.

Von links nach rechts, von oben nach unten:
Brüstung mit Maschen, Ansicht der Treppenstütze,
Rückansicht, Blick auf die Treppe.
Rechts: Gekröpftes Spannband.

De gauche à droite, du haut vers le bas:
Rambarde, vue du tablier, vue partielle
avec la rivière, vue de l'escalier.
À droite: Pont caténaire.

CENTRAL PARK ENTRANCE GATEWAY PROJECT,
MANCHESTER, UNITED KINGDOM

GIFFORD WITH AUKETT FITZROY ROBINSON

www.gifford.uk.com, www.aukett.com

Type: Post tensioned finback, **Length:** 131.8 m, **Contractor:** Balfour Beatty, Bilfinger Berger (Joint Venture), **Completion:** 2006, **Photos:** Courtesy of Gifford, Aukett Fitzroy Robinson/London (260 b. l., 261).

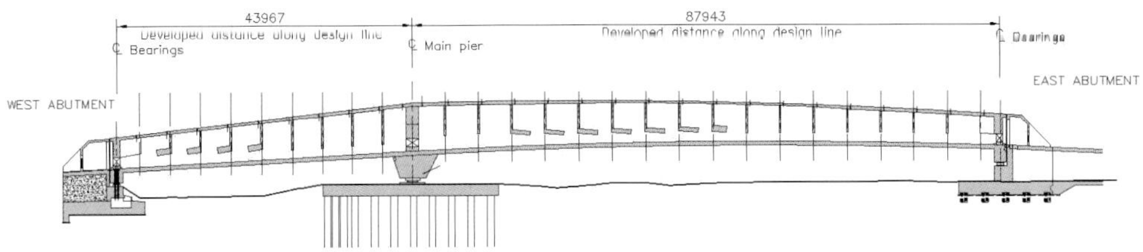

Left: Aerial view. Links: Luftaufnahme. À gauche: Vue aérienne. | Right: Sectional elevation. Rechts: Ansicht der Aufteilung. À droite: Vue en coupe.

Part of the Metrolink phase 3 project passes through the Central Business Park and crosses an existing railway. A post-tensioned concrete finback bridge has been provided to carry the Metrolink tracks over the existing rail route. The bridge was constructed in a position alongside to and roughly parallel to the railway. When nearly completed, the bridge was rotated 21 degrees to reach its final position. The site constraints resulted in a reversed curve horizontal alignment resulting in significant torsion in the deck. The solution was a box section finback bridge with varying height and wall thickness with internal tendons in the deck slab and external tendons inside the fin.

Ein Teil des Projektes Metrolink Phase 3 führt durch den Stadtpark und kreuzt bestehende Schienenwege. Eine nachgespannte Flossenbrücke wurde dafür vorgesehen, die Metrolink-Stränge über die existierenden Gleise zu führen. Die Brücke wurde entlang der bestehenden Gleise errichtet und im fast fertigen Zustand um 21 Grad zu ihrer jetzigen Position gedreht. Die Umstände am Bauort resultierten in einer S-Kurve mit horizontaler Angleichung, die eine erhebliche Torsion der Brückenoberfläche zur Folge hatte. Die Lösung bot eine in Kastensegmente unterteilte Flossenbrücke mit variierender Höhe und Wanddicke, inneren Verspannungen in der Fahrbahnplatte und äußeren Verspannungen in der Flosse.

Le tracé de la phase 3 du projet Metrolink de Manchester traverse une zone industrielle et passe au-dessus d'une ligne de chemin de fer préexistante. Il intègre notamment un pont « queue de poisson » en béton. L'ouvrage est tout d'abord parallèle à la voie ferrée, puis la franchit en formant un angle de seulement vingt-et-un degrés. Cette torsion et ces formes sinueuses sont le résultat des contraintes imposées par le site. La solution retenue consiste en des caissons en béton de taille et d'épaisseur variables, renforcés à l'intérieur au niveau du tablier, et à l'extérieur au niveau de la « queue ».

From left to right, from above to below:
Existing railroad, cantilever, back span prior to rotation, completed structure.
Right: Completed structure with adjacent station canopy.

Von links nach rechts, von oben nach unten:
Bestehende Schienenwege, Auskragung, aufgespannte Konstruktion vor der Drehung, fertige Konstruktion.
Rechts: Fertige Struktur und Dach der benachbarten Station.

De gauche à droite, du haut vers le bas:
Voie de chemin de fer, rambarde en porte-à-faux, vue d'ensemble, tablier vu d'en bas.
À droite: Vue aérienne figurant le pont et la gare voisine.

RIVER HULL FOOTBRIDGE,
RIVER HULL, UNITED KINGDOM

MCDOWELL+BENEDETTI

www.mcdowellbenedetti.com

Type: Swing bridge, **Length:** 57 m, cantilever 40 m, **Structural engineers:** Alan Baxter & Associates, **Project manager:** Arup, **M&E engineers:** Qualter Hall&Co, **Lighting:** Sutton Vane Ass, **Landscape:** Whitelaw Turkington, **Artist:** Nayan Kulkarn, **Client:** Yorkshire Forward, **Completion:** 2010, **Renderings:** McDowell+Benedetti/London.

Left: Bird's-eye view. Links: Luftaufnahme. À gauche: Le pont vu de haut. | Right: Structural skeleton. Rechts: Strukturskelett. À droite: « Squelette » de la passerelle.

This mobile bridge across the River Hull allows pedestrians to remain on it while it moves. Its unusual form and the new public spaces at either side create a new connection and a memorable place to visit. It provides two generous pedestrian/cycle routes – one gently sloping, the other a more direct stepped route. From the braced steel drum, which acts as a counterbalance and houses a café, the cantilevered steel spine arches and curves over the river to allow smaller boats to pass underneath. When required, the bridge rotates on an electrical drive mechanism to open for larger boats, which is a striking event enhanced by an art installation and by special lighting.

Diese bewegliche Brücke über den Fluss Hull erlaubt Passanten den Aufenthalt während sie sich bewegt. Ihre ungewöhnliche Form und neu entstandene öffentliche Räume an den Ufern schaffen eine Verbindung und einen unvergesslichen Ort. Die Brücke hat zwei Fußgänger- und Radwege, wovon der eine im sanften Bogen verläuft, während der andere direkter und abgestuft ist. Von der als Gegengewicht fungierenden Stahltrommel, die ein Cafe beherbergt, wölbt und krümmt sich der ausladende Stahlrücken über den Fluss und ermöglicht kleinen Booten die Durchfahrt. Für größere Boote dreht sich die Brücke mit Hilfe eines elektrischen Mechanismus – dank der Beleuchtung und einer Kunstinstallation ein besonderes Ereignis.

Cette passerelle mobile présente la particularité d'être accessible même lorsqu'elle est en mouvement. Sa forme inhabituelle, les espaces qui y sont aménagés et l'installation sonore et visuelle activée lorsqu'elle est en position ouverte font qu'elle est très appréciée du public. Le tablier en acier porte deux chaussées, la première en pente douce, la seconde plus directe. Positionné à une distance suffisante au-dessus de la rivière pour permettre le passage des canots, il a pour contrepoids un grand tambour également en acier qui abrite un café. L'ensemble, actionné par un moteur électrique, tourne sur un axe situé au centre du café pour laisser passer les grandes embarcations.

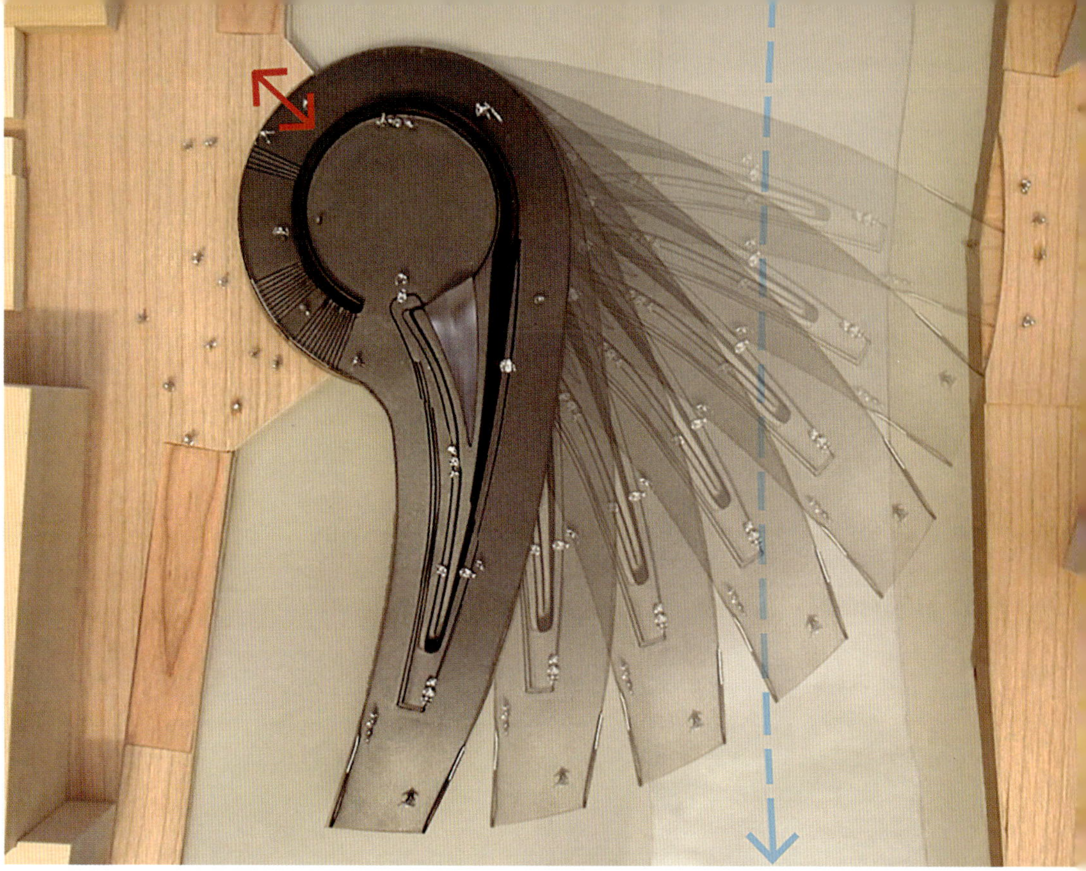

From left to right, from above to below:
Model view from east, movement of the bridge,
view from north, view from west.
Right: View at night.

Von links nach rechts, von oben nach unten:
Modellansicht von Osten, Brückenbewegung,
Modellansicht von Norden, Modellansicht von Westen,
Rechts: Nachtansicht.

De gauche à droite, du haut vers le bas:
Maquette vue de l'est, passerelle en mouvement,
maquette vue du nord et de l'ouest.
À droite: La passerelle la nuit.

FABIAN WAY BRIDGE,
SWANSEA, UNITED KINGDOM

STUDIO BEDNARSKI LTD.

www.studio-bednarski.com

Type: Cable-stayed bridge, **Length:** 75 m, **Structural engineers:** Parsons Brinckerhoff + Flint&Neill Partnership, **Completion:** 2007, **Photos:** Studio Bednarski Ltd (266), CM Bednarski London (268), David Morrell, courtesy of Lighting Technology Projects Ltd. (269).

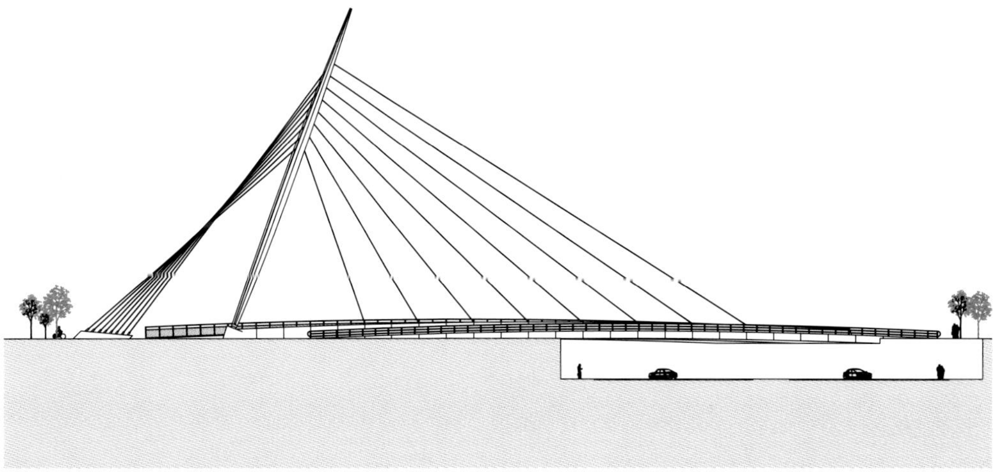

Left: Perspective. Links: Perspektive. À gauche: Vue du pont sur l'autoroute. | Right: Side elevation. Rechts: Seitenansicht. À droite: Vue en élévation.

This bridge provides urgently needed pedestrian, cyclist and bus routes across the Fabian Way (A483) near Port Tennant for the city of Swansea. Its concept and detail were defined by its location and the very oblique angle across Fabian Way. It is a cable-stayed solution with a unique forward inclined leaf-like mast, which conceals the cable attachments. A spine beam separates pedestrian and bus lanes. The twisted arrangement of its back stays is a further feature that complements the uniqueness of the structure. At night the bridge is richly illuminated constituting a landmark for the city.

Diese Brücke stellt dringend benötigte Verbindungsrouten für Fußgänger, Radfahrer und Busse über den Fabian Way (A483) in der Nähe des Hafens Tennant für Swansea bereit. Das Konzept und die Feinheiten der Brücke werden durch die Besonderheit ihrer Lage und den schrägen Winkel über den Fabian Way bestimmt. Es handelt sich um eine Schrägseilbrücke mit vorwärts geneigtem blattähnlichem Pylon, der die Kabelbefestigungen verbirgt. Ein Grat unterteilt Fußgänger und Busspuren. Die verdrillte Anordnung der Stützstreben ist ein weiteres, die Einzigartigkeit der Konstruktion ergänzendes Merkmal. In der Nacht wird die Brücke reichlich und farbenfroh beleuchtet und wird so zum Wahrzeichen der Stadt.

Il était plus qu'urgent de construire un pont pour bus, piétons et cyclistes au-dessus de l'autoroute A483 (Fabian Way) qui dessert Port Tennant, à Swansea. Son tracé est parfaitement inhabituel puisqu'il traverse la voie en question non pas perpendiculairement, mais en oblique. Il s'agit d'un pont suspendu, les haubans étant attachés à un pylône unique en position inclinée. Une glissière sépare la voie des bus de celle réservée aux piétons et cyclistes. La disposition torsadée des haubans constitue une autre caractéristique de l'ouvrage. La nuit, un dispositif d'éclairage très élaboré fait du pont une sculpture lumineuse qui domine la ville.

From left to right, from above to below:
View of leaf mast, view along bridge,
deck, detail of twisted stays.
Right: General view at night.

Von links nach rechts, von oben nach unten:
Blick auf den blattförmigen Pylon, Blick entlang der Brücke,
Fahrbahn, Detailansicht der verdrillten Verstrebungen.
Rechts: Gesamtansicht bei Nacht.

De gauche à droite, du haut vers le bas:
Sommet du pylône, vue du tablier, voies menant
au pont, haubans torsadés.
À droite: Vue d'ensemble la nuit.

LEONARD P. ZAKIM BUNKER HILL BRIDGE,
BOSTON, MA, USA

ROSALES + PARTNERS

www.rosalespartners.com

Type: Cable-stayed bridge, **Length:** 449.58 m, **Structural engineers:** Christian Menn, Parsons Brinckerhoff, HNTB, **Completion:** 2003, **Photos:** John Wolf (272), Andy Ryan (270, 273).

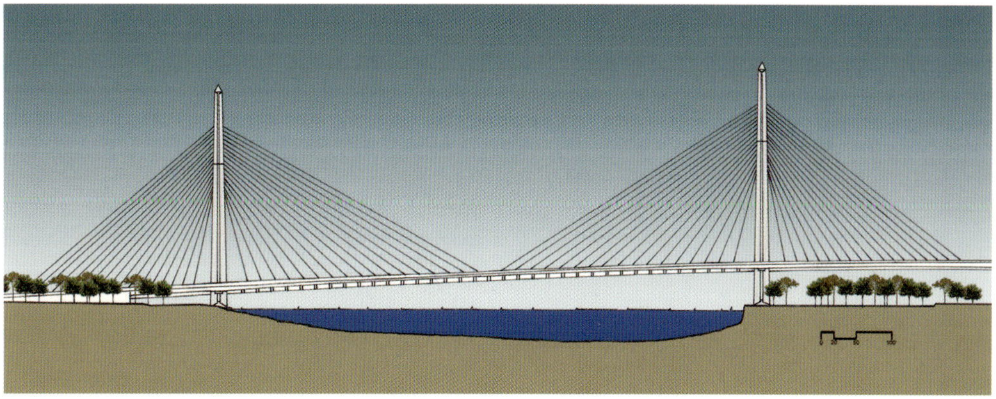

Left: View from TD Banknorth Garden. Links: Blick vom TD Banknorth Garden. À gauche: Le pont vu du TD Banknorth Garden. | Right: Section. Rechts: Längsschnitt. À droite: Vue en coupe.

The Zakim Bridge was part of the Central Artery/Tunnel Project. The 1,475-foot structure features twin inverted Y-shaped towers and a 745-foot main span. It carries ten lanes of traffic – eight interstate lanes passing through the towers and two local traffic lanes cantilevered off the east side, outside the towers. The towers resemble the shape of the historic Bunker Hill Monument in neighboring Charlestown, while the bridge's overall form echoes the shape of the sailboats that navigate the inner Charles River Basin and Boston Harbor. The bridge is a striking addition to the Boston skyline and a new symbol of the city.

Die Zakim-Brücke entstand als Teil des Central Artery/Tunnel Project. Die 1.475 Fuß umfassende Struktur weist zwei gegabelte Pylone und eine Spanne von 745 Fuß auf. Zehn Fahrspuren haben auf ihr Platz, wovon acht Bundesstraßen sind und durch die Pylone führen und zwei Nahverkerhrsstraßen an der östlichen Außenseite der Pylone auskragen. Die Form der Pylone erinnert an das Bunker Hill Monument im benachbarten Charlestown während die Gestalt der Gesamtkonstruktion eher der Kontur der Segelboote entspricht, die das Charles River Basin und den Bostoner Hafen ansteuern. Die Brücke ist eine augenfällige Erweiterung der Bostoner Skyline und ein neues Wahrzeichen für die Stadt.

Ce pont a été réalisé en même temps que le tunnel et l'artère centrale de Boston. D'une longueur totale de 450 mètres, il est supporté par deux piles en forme de « Y » renversé espacées de 227 mètres. Une autoroute à deux fois cinq voies passe entre les jambes des piles, tandis qu'une route de desserte locale est en porte-à-faux à l'extérieur sur le côté est. Les piles rappellent le monument de Bunker Hill situé non loin d'ici, à Charlestown, tandis que la forme générale du pont évoque les voiliers qui naviguent dans le port et sur le Charles River Basin. Spectaculaire et visible de loin, l'ouvrage est le nouveau symbole de la ville de Boston.

From left to right, from above to below:
Y-shaped pylons at night, pylon during daytime, bridge as dominating part of the skyline.
Right: View from Charlestown.

Von links nach rechts, von oben nach unten:
Gegabelte Pylone bei Nacht, Pylon bei Tageslicht, Brücke als dominierender Teil der Skyline.
Rechts: Blick von Charlestown.

De gauche à droite, du haut vers le bas:
Les piles en forme de « Y » la nuit et le jour, le pont dominant la silhouette de la ville.
À droite: Le pont vu de Charlestown.

SKYWALK GRAND CANYON WEST,
GRAND CANYON, AZ, USA

MRJ ARCHITECTS, INGENIEURBÜRO LOCHSA ENGINEERING, APCO CONSTRUCTION

www.destinationgrandcanyon.com

Type: Anchored cantilever bridge, **Length:** 47 m, **Structural engineers glass basement:** SG Kinon Porz (Saint-Gobain Deutsche Glas), **Development and production glass breast:** Döring Glas (Saint-Gobain Deutsche Glas), **Completion:** 2007, **Photos:** Grand Canyon West.

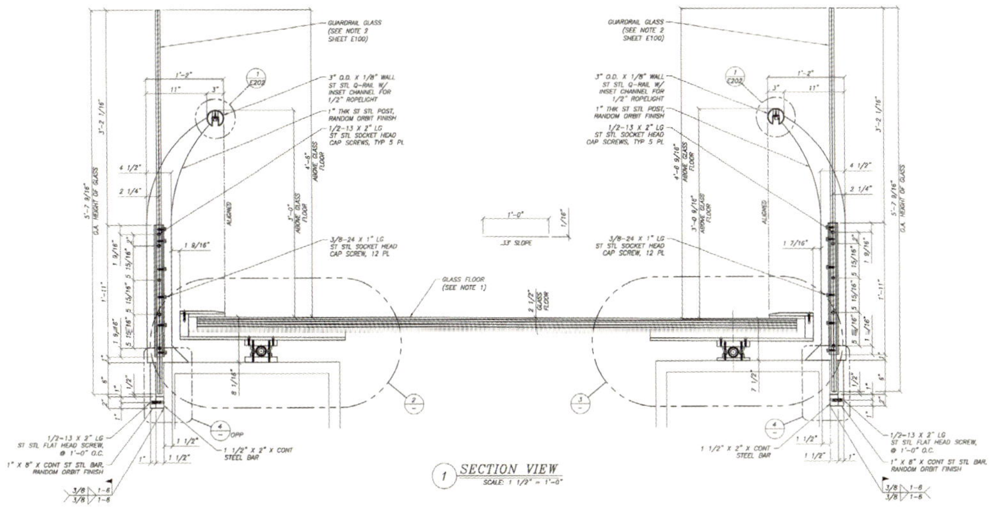

Left: Skywalk and Grand Canyon. Links: Skywalk und Grand Canyon. À gauche: « Skywalk » et Grand Canyon. | Right: Section. Rechts: Schnitt. À droite: Vue en coupe.

A horseshoe-shaped balcony, a bridge leading back to the same bank presents a view of the Grand Canyon. The structure rests more than a kilometer above the Colorado River, jutting 21 meters out into the abyss. The floor of the 482-ton steel-and-glass structure consists of 20 square and 26 wedge-shaped compound safety glass elements. The glass floor, sealed with a special laminate, has an area of 147 square meters and can support up to 120 people at a time, holding a weight of 750 kilos per square meter. A 100-meter long curved glass railing, made from extra-transparent glass panels, completes the illusion of flying over the canyon for those standing on the balcony.

Eine Brücke, die an das Ausgangsufer zurückführt, ein hufeisenförmig ausgebildeter Balkon, erschließt den Blick in den Grand Canyon. Mehr als einen Kilometer oberhalb des Colorado River kragt die Konstruktion 21 Meter in den Abgrund hinaus. Der Boden der 482 Tonnen schweren Stahl- und Glaskonstruktion umfasst 20 rechteckige und 26 trapezförmige Sicherheitsglaselemente. Der 147 Quadratmeter große, laminierte Glasboden trägt bis zu 120 Personen gleichzeitig und hält je Quadratmeter einer Last von 750 Kilogrammm stand. Die 100 Meter lange, gebogene Glasbrüstung aus besonders transparentem Glas macht die Illusion, man fliege über den Canyon, perfekt.

Une plate-forme en fer à cheval a été récemment construite au sommet du Grand Canyon du Colorado. Elle s'avance de vingt-et-un mètres au-dessus du vide, avec le fleuve coulant un millier de mètres en contrebas. Cette structure qui pèse plus de quatre cents tonnes se compose d'une ceinture en acier supportant des plaques de verre Sécurit assemblées en utilisant une colle spéciale pour former une plate-forme de 147 mètres carrés pouvant accueillir jusqu'à 120 personnes à la fois (soit un poids de 750 kilos au mètre carré). La rambarde incurvée d'une centaine de mètres de long a elle aussi été réalisée en verre de manière à parfaire l'illusion qu'ont les visiteurs de planer au-dessus du vide.

From left to right, from above to below:
Cantilever, view of the Grand Canyon, Hualapai
Native American in front of the Skywalk.
Right: Bird's-eye view.

Von links nach rechts, von oben nach unten:
Auskragung, Blick über den
Grand Canyon, Hualapai-Indianer
vor dem Skywalk, Ansicht von unten.
Rechts: Luftaufnahme.

De gauche à droite, du haut vers le bas:
Plate-forme en porte-à-faux, « Skywalk » et Grand Canyon,
Indien Hualapai avec la plate-forme et
le canyon à l'arrière plan.
À droite: Plate-forme vue d'en bas.

LIBERTY BRIDGE,
GREENVILLE, SC, USA

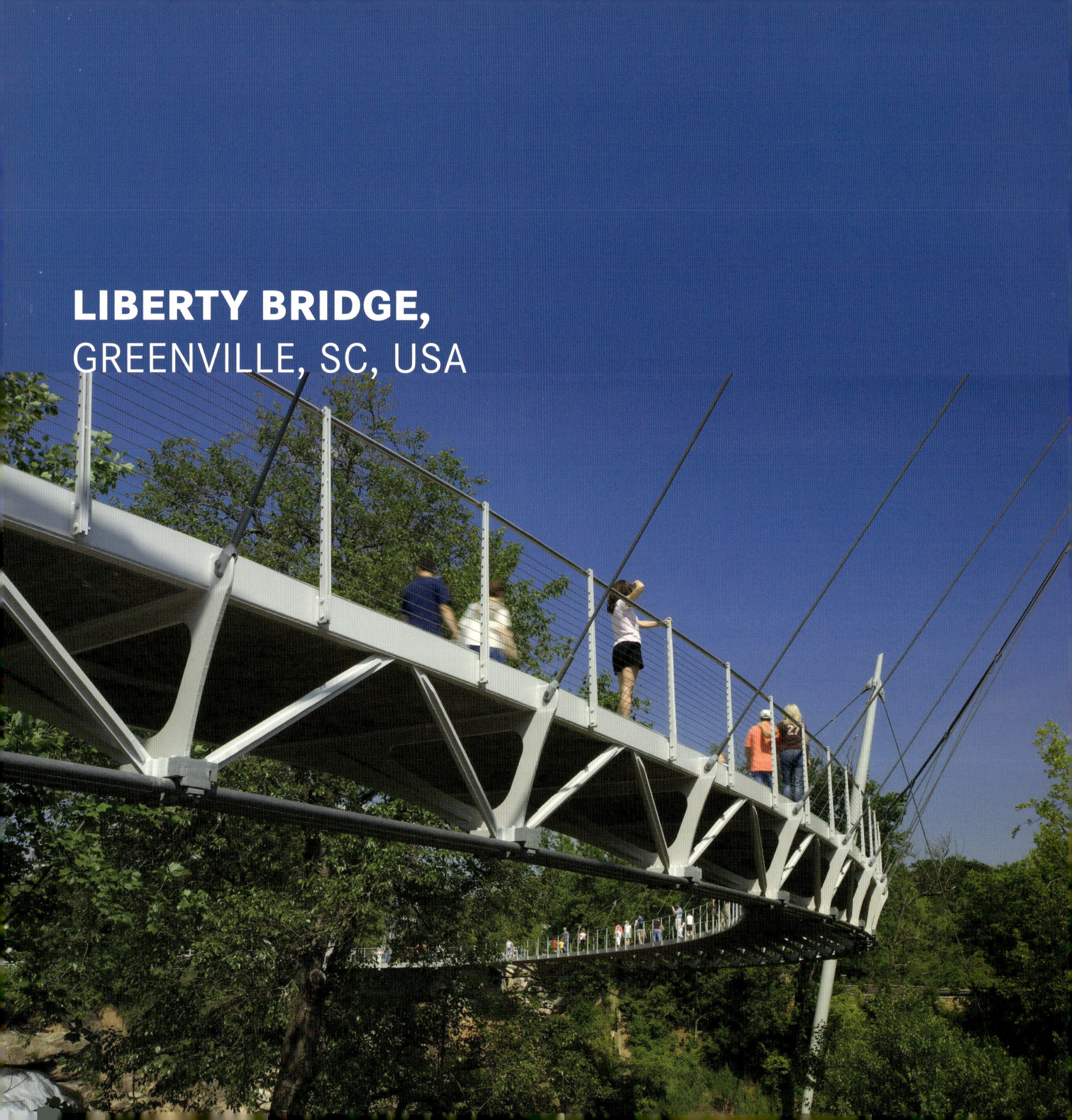

ROSALES + PARTNERS

www.rosalespartners.com

Type: Suspension bridge, **Length:** 115.82 m, **Structural engineers:** Schlaich Bergermann and Partner, **Completion:** 2004, **Photos:** Kun Zhang - InfiniDimension.

Left: Combined bridge and viewing platform. Links: Brücke und Aussichtsplattform in einem. À gauche: Passerelle panoramique. | Right: Sketch. Rechts: Skizze. À droite: Esquisse.

Greenville's downtown is divided by a forested valley park containing the falls of the Reedy River. The Liberty Bridge is located in the immediate vicinity of these waterfalls. The bridge has a clear span over the river that curves away from the falls, providing visitors with an aerial platform from which to admire the cascading water. The bridge gently slopes into the ravine and is supported by widely visible twin inclined towers and a single suspension cable, allowing unobstructed views. With a total length of approximately 380 feet, the structure appears to float over the landscape, offering a stunning view of the park.

Die Innenstadt Greenville's wird von einem bewaldeten Park geteilt, in welchem sich die Wasserfälle des Reedy River befinden. Direkt an den Wasserfällen liegt die Liberty Bridge. Ihre freie Spannweite krümmt sich fort von den Wasserfällen und bildet eine hohe Plattform, die es Besuchern ermöglicht, das herabstürzende Wasser zu bewundern. Die Brücke windet sich sanft durch die Schlucht und wird von zwei weithin sichtbaren, geneigten Pylonen und einem einzelnen Kabel getragen. Die Konstruktion scheint mit ihrer Länge von fast 380 Fuß über der Landschaft zu schweben und ermöglicht atemberaubende Ausblicke auf den Park.

Les rives boisées de la rivière Reedy forment un parc au cœur de l'agglomération de Greenville. Le Liberty Bridge, long d'environ 115 mètres, est une passerelle de forme incurvée construite juste en aval des chutes de la rivière afin de permettre au public d'apprécier le spectacle naturel. L'ouvrage est suspendu par des haubans à deux piliers inclinés visibles de loin et reliés par un seul câble. Grâce à cette structure ultra-légère qui donne l'impression que la passerelle flotte en l'air, les promeneurs bénéficient d'un panorama complet sur le paysage environnant.

From left to right, from above to below:
View from river bed, pedestrian route through the park, light bridge structure with waterfalls in the back.
Right: Night illumination.

Von links nach rechts, von oben nach unten:
Blick vom Flußbett, Fußgängerweg durch den Park, leichter Brückenrahmen vor den Wasserfällen.
Rechts: Nachtbeleuchtung.

De gauche à droite, du haut vers le bas:
Vue de la passerelle et des chutes, piétons sur la passerelle, autre vue de la passerelle et des chutes.
À droite: Passerelle illuminée la nuit.

SUNDIAL BRIDGE AT TURTLE BAY,

REDDING, CA, USA

SANTIAGO CALATRAVA ARCHITECT & ENGINEER

www.calatrava.com

Type: Cable-stayed bridge, **Length:** 700 ft., **Client:** City of Redding, McConnell Foundation, **Completion:** 2004, **Photos:** Alan Karchmer / Architectural Photographer, **Plans:** Santiago Calatrava.

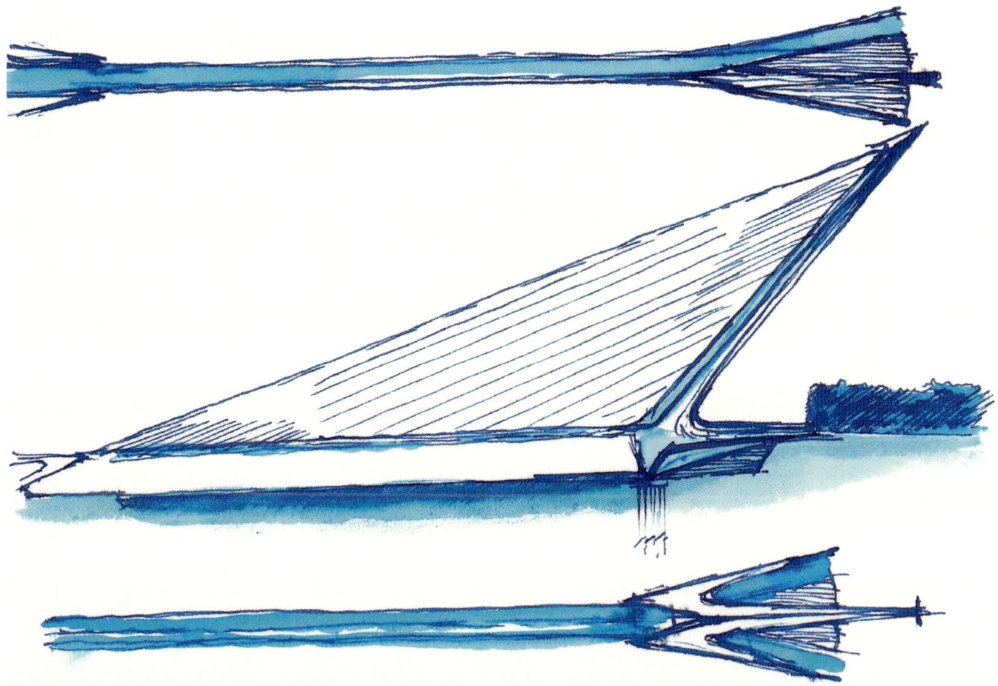

Left: View along the river. Links: Luftaufnahme. À gauche: Vue de la passerelle et du fleuve. | Right: Sketches. Rechts: Entwurfszeichnungen. À droite: Esquisses.

Located on the Sacramento River in Northern California, Turtle Bay Exploration Park is a campus surrounding the Turtle Bay Museum, which features exhibits exploring the relationship between humans and nature. In 1997, Santiago Calatrava was commissioned by the City of Redding, supported by the McConnell Foundation with the construction of this pedestrian bridge, linking the park's museum building to its arboretum. The bridge also serves as a new downtown entrance for Redding's Sacramento River trail system. To account for the park's mission and for the delicate nature of the site the bridge was designed without pillars in the water.

Der Turtle Bay Exploration Park, ein Campus am Sacramento River im Norden Kaliforniens, ist um das Turtle Bay Museum angelegt. Hier finden Ausstellungen zur Beziehung von Mensch und Natur statt. 1997 wurde Santiago Calatrava von der Stadt Redding mit Unterstützung der McConnell Foundation mit dem Bau dieser Fußgängerbrücke betraut, die das Parkmuseum mit dem Arboretum verbindet. Sie ist darüber hinaus der neue Zugang Reddings zum Wandergebiet am Sacramento River. Die Brücke wurde ohne Stützpfeiler im Wasser gestaltet, um sowohl der Intention des Parks nachzukommen, Menschen für die Natur zu begeistern als auch um die sensible Naturlandschaft zu bewahren.

Le Turtle Bay Museum, consacré aux relations entre l'Homme et la nature, se complète d'un « parc d'exploration » qui longe le fleuve Sacramento. En 1997, les autorités locales et la fondation McConnell ont chargé Santiago Calatrava de construire une passerelle sur le fleuve pour relier directement le parc à un arboretum, et pour rendre les sentiers de randonnée de la région plus accessibles aux habitants du centre ville. La passerelle, qui offre aux promeneurs des vues magnifiques sur le paysage, franchit le fleuve sans aucune pile dans l'eau de manière à respecter la mission du parc en minimisant l'impact sur l'environnement.

From left to right, from above to below:
View from deck, worms-eye view of pylon,
view from river shore, detail of framework.
Right: View along the deck.

Von links nach rechts, von oben nach unten:
Blick vom Deck, Pylon von unten betrachtet,
Blick vom Flussufer, Fachwerkstruktur.
Rechts: Blick über das Deck.

De gauche à droite, du haut vers le bas:
Vue dans l'axe de la passerelle au crépuscule,
pylône vu d'en dessous, passerelle vue de la rive,
détail de la structure du tablier.
À droite: Vue du tablier et du pylône.

MAPLE AVENUE BRIDGE,
REDMOND, OR, USA

JIRI STRASKY WITH OBEC CONSULTING ENGINEERS, EUGENE

www.shp.eu, www.obec.com

Type: Arched bridge, **Length:** 238 m, **Completion:** 2007, **Photos:** Jiri Strasky.

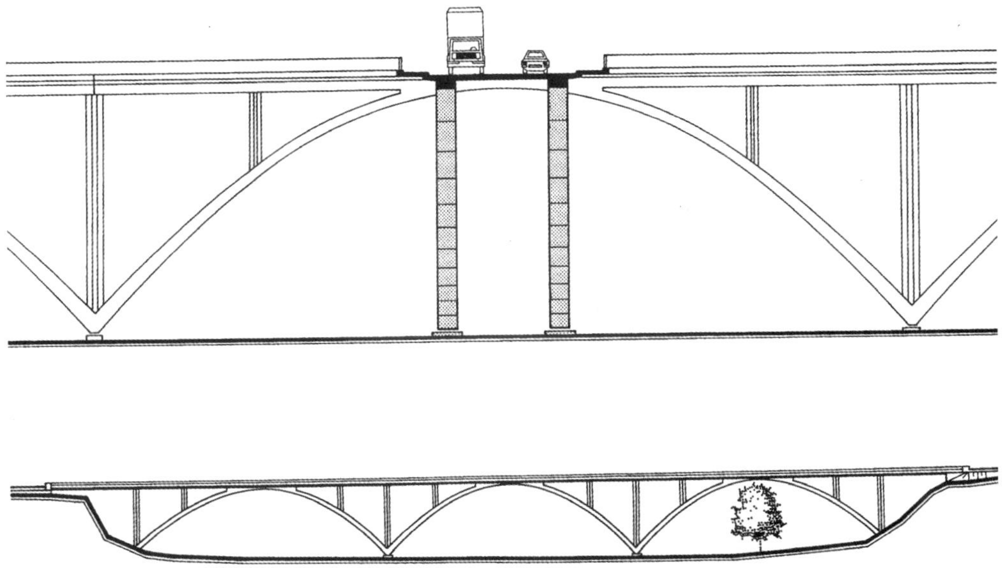

Left: View from below. Links: Unterseite. À gauche: Pont vu d'en dessous. | Right: Cross section and elevation. Rechts: Querschnitt und Ansicht. À droite: Vue en coupe et élévation.

The Maple Avenue Bridge provides an east-west link for the City of Redmond across the Dry Canyon. Dividing the city Dry Canyon is a unique landscape which provides recreational areas for leisure activities. From the beginning it was clear that a bridge consisting of three continuous arch spans and two approach spans would be the most aesthetically pleasing solution. However, the bridge also had to be built within an affordable budget. Therefore the designer developed a very clear and simple structure without deck diaphragms and arch bracing.

Die Maple Avenue-Brücke bietet eine Ost-West-Verknüpfung über den Dry Canyon für die Stadt Redmond. Dry Canyon ist eine landschaftliche Besonderheit, welche die Stadt unterteilt und Freiflächen zur Erholung bietet. Von Anfang an war klar, dass eine Brücke mit drei fortlaufenden Bogenspannen und zwei Spannen an den Auffahrten die ästhetisch ansprechendste Lösung wäre. Allerdings musste die Brücke auch im finanzierbaren Rahmen bleiben. Aus diesem Grund wurde eine klare und einfache Struktur entwickelt, in der Zwischenfahrbahndecken und querlaufende Träger vermieden wurden.

Ce pont orienté est/ouest traverse le Dry Canyon qui partage la ville de Redmond en une moitié nord et une moitié sud. Le canyon abritant des espaces récréatifs naturels appréciés des citadins, les architectes ont opté pour une structure esthétique composée de trois arches successives. Le budget disponible étant d'autre part restreint, ils ont réalisé un ouvrage simple et aux lignes parfaitement claires, sans renforts transversaux ni couches intermédiaires au niveau du tablier.

From left to right, from above to below:
Arch bent elevation, view from below,
piers, detail view from below.
Right: Arch spans.

Von links nach rechts, von oben nach unten:
Gekrümmter Bogen, Ansicht von unten,
Stützen, Detailansicht von unten.
Rechts: Brückenbögen.

De gauche à droite, du haut vers le bas:
Départ de deux arches, pont vu d'en dessous,
arches et tablier, détail du tablier vu d'en dessous.
À droite: Pont vu dans l'axe du canyon.

T. EVANS WYCKOFF MEMORIAL BRIDGE,

SEATTLE, WA, USA

SRG PARTNERSHIP

www.srgpartnership.com

Type: Hoop truss bridge, **Length:** 104 m, **Structural engineers:** Magnusson Klemencic Associates, **Completion:** 2008, **Photos:** Lara Swimmer, Seattle.

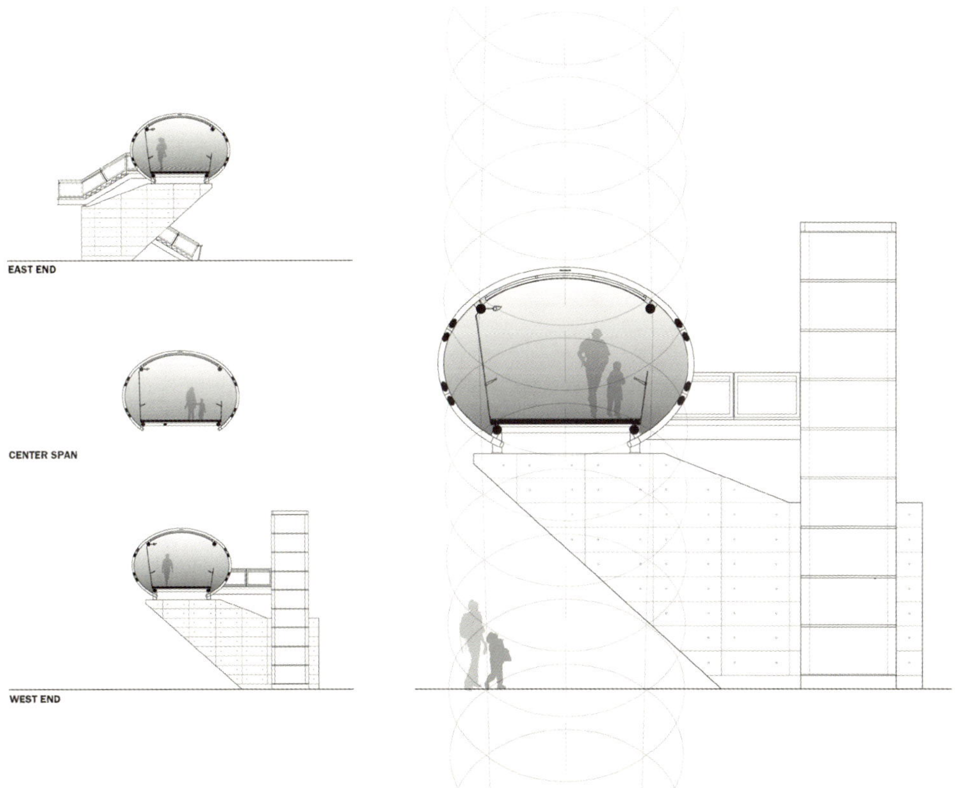

Left: View of the bridge west end. Links: Blick auf das westliche Brückenende und Richtung Museum. À gauche: Vue d'une extrémité de la passerelle avec le musée à l'arrière-plan. | Right: Cross-section diagrams of the bridge truss. Rechts: Querschnitt der fachwerkröhre. À droite: Différentes vues en coupe.

The growing collection of one of the largest air and space museums in the world made additional exhibit space on the opposite side of a busy traffic arterial necessary. A safe and convenient pedestrian crossing for museum visitors maintaining the spirit of the museum was subsequently built. The design was inspired by the phenomenon of a contrail, a stream of crystallized vapor created in a plane's wake. The metaphor is carried out in the bridge's unusual tube truss design, a 340' long open structure surrounding an inner glass enclosure. The unique structure does not rely on conventional truss webs, but instead distributes the vertical shear in the bridge structure through a matrix of curving steel pipes.

Die wachsende Sammlung eines der größten Luftfahrtmuseen der Welt machte eine Erweiterung der Ausstellungsflächen jenseits einer verkehrsreichen Straße nötig. Eine sichere Überquerung, die den Besucher zugleich in der Sphäre der Ausstellung hält wurde errichtet. Das Design beruht auf Kondensstreifen, kristallisiertem Wasserdampf den Flugzeuge hinter sich herziehen. Diese Metapher ist in der ungewöhnlichen Fachwerkröhre, die auf 340 Fuß Länge eine Innere Glashaut umschließt, aufgegriffen. Es handelt sich nicht um einfaches Fachwerk, sondern greift in dem Muster der gebogenen Röhren die vertikalen Schubkräfte auf.

L'accroissement continuel des collections d'un des plus grands musées au monde consacré à l'aviation a rendu nécessaire de construire une annexe, reliée au bâtiment principal par une passerelle enjambant une grande artère. La forme générale de cet ouvrage d'une centaine de mètres de long évoque les traînées de vapeur cristallisées laissées par un avion à réaction volant à haute altitude. Il s'agit d'une structure tubulaire constituée d'un entre-lac de tubes incurvés formant une enveloppe autour d'une passerelle vitrée.

From left to right, from above to below:
View from below, detail of curving steel pipes , bridge at night.
Right: View of the bridge interior.

Von links nach rechts, von oben nach unten:
Blick von unten, Detailansicht der gebogenen Stahlrohre, Nachtaufnahme.
Rechts: Blicke ins Innere der Brücke.

De gauche à droite, du haut vers le bas:
Passerelle vue d'en bas, gros plan sur la passerelle
vue du sud, passerelle au crépuscule.
À droite: Intérieur de la passerelle.

CHIHULY BRIDGE OF GLASS,
TACOMA, WA, USA

ANDERSSON • WISE ARCHITECTS

www.anderssonwise.com

Type: Span bridge, **Length:** 482 ft., **Glass artist:** Chihuly Studio, **Completion:** 2002, **Photos:** Tim Hursley.

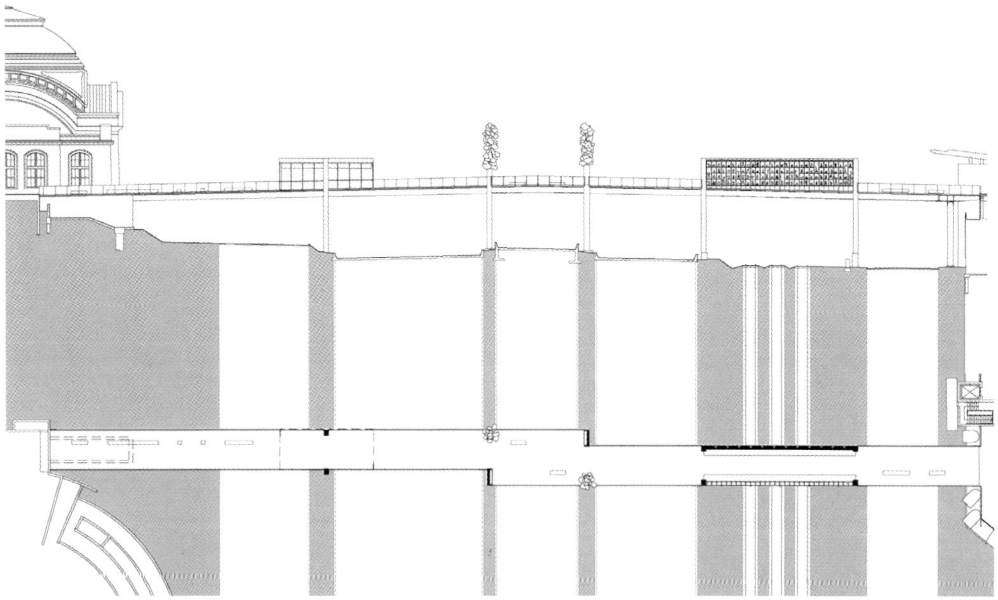

Left: View north of the Venetian Wall and Crystal Gate from rail lines. Links: Blick nach Norden zur Venezianischen Wand und zum Kristalltor. À gauche: Face nord de la passerelle, avec le Mur vénitien et la Porte de cristal, vue de la voie de chemin de fer. | Right: Section and plan. Rechts: Grundriss und Aufriss. À droite: Élévation et plan.

More than a crucial pedestrian connection linking the Museum of Glass with the developing historic/cultural district, the Bridge of Glass is an energized public experience. It was essential to allow visitors to pause, sit, look and be enveloped in Chihuly's art. Site-specific pavilions respond to the different types of objects featured, creating different worlds of Chihuly glass. The bridge grows from its material context of concrete and steel. Massive concrete supports and the painted steel superstructure have a powerful yet neutral presence that is in contrast to Chihuly's artwork. The idea was not to compete for attention but to create a prosaic counterpoint to the poetry of glass.

Die Brücke aus Glas ist nicht nur eine wichtige Anbindung des Museum of Glass mit dem wachsenden historisch-kulturellen Viertel, sondern ein öffentliches Erlebnis. Wichtigstes Anliegen war es, den Besuchern einen Ort zum Pausieren zu schaffen, der sie mit Chihulys Kunst umschließt. Ortsspezifische Pavillons greifen die ausgestellten Objekte auf und kreieren verschiedene Welten aus Glas. Die Brücke erwächst aus ihrem Kontext aus Beton und Stahl. Die massiven Betonstützen und der Stahloberbau erzeugen eine starke und doch neutrale Präsenz, die im Kontrast zu den Glasarbeiten steht. Es ging hier nicht darum, um Aufmerksamkeit zu buhlen, sondern einen Kontrast zur Poesie des Glases zu schaffen.

Plus qu'une simple passerelle entre le Museum of Glass et le centre ville historique de Tacoma, cet ouvrage est une véritable expérience culturelle mise à la disposition du public, qui peut s'y asseoir pour apprécier à son aise des œuvres en verre créées par Dale Chihuly et exposées dans divers pavillons. La passerelle se compose d'une structure porteuse en béton et plaques d'acier peintes d'un aspect général assez neutre, qui s'inscrit en contraste par rapport aux œuvres d'art multicolores. Les architectes se sont efforcés de créer un cadre prosaïque apte à mettre en valeur la poésie du verre.

From left to right, from above to below:
Crystal Gate, Venetian Wall,
view East from Seaform Pavilion.
Right: View from Seaform Pavilion of
Washington State History Museum.

Von links nach rechts, von oben nach unten:
Kristalltor, Venezianische Wand,
Blick nach Osten vom Meeresformen-Pavillon.
Rechts: Washington State History Museum.

De gauche à droite, du haut vers le bas:
Porte de cristal, mur vénitien, vue du pavillon
« Seaform » en direction de l'est.
À droite: Vue du pavillon « Seaform » en direction
du Washington State History Museum.

WOODROW WILSON MEMORIAL BRIDGE,

WASHINGTON, D.C., USA

ROSALES + PARTNERS

www.rosalespartners.com

Type: Steel haunched girders & V-shaped concrete piers / Baseline bridge, **Length:** 1,828.8 m, **Structural engineers:** Parsons Transportation Group, **Completion:** 2008, **Photos:** Alan Karchmer Architectural Photographer, Washington, D.C.

Left: View from shoreline. Links: Blick vom Ufer. À gauche: Le pont vu de la rive. | Right: Diagram of opening process. Rechts: Diagramm der Brückenöffnung. À droite: Pont en position ouverte et fermée.

The new Woodrow Wilson Memorial Bridge carries Interstate 95 and the Capital Beltway across the Potomac River. Connecting Alexandria, Virginia, with Prince George's County, Maryland, it is one of Washington's key national monuments. The mobile bridge design incorporates 18 sets of V-piers that resemble seagulls in flight, giving the twin spans a light and open appearance. The piers form an arcade reminiscent of traditional Washington bridges. The V-piers support concrete-covered steel girders that balance the load subjecting the piers to minimal dead load bending moments and protecting the foundations from horizontal arch thrust.

Die Woodrow Wilson Memorial Bridge führt die Interstate 95 und den Capital Beltway über den Potomac River. Sie verbindet Alexandria mit Prince Georges County und ist Teil der zentralen Washingtoner nationalen Denkmäler. Der Entwurf dieser beweglichen Brücke umfasst 18 Reihen V-förmiger Stützpfeiler, die fliegenden Möwen gleichen und der Spannweite eine leichte und offene Erscheinung verleihen. Die Arkade der Stützpfeiler erinnert an traditionelle Washingtoner Brücken. Auf den Stützpfeilern liegen betonummantelte Stahlträger, welche die Last ausgleichen, um die Pfeiler nur minimaler Biegebeanspruchung durch das ruhende Gewicht auszusetzen und die Fundamente vor horizontalem Gewölbedruck zu bewahren.

Ce pont permet à l'interstate 95 et au périphérique de Washington de traverser le Potomac. Reliant Alexandria, en Virginie, au Prince George's County, au Maryland, l'ouvrage fait partie des monuments nationaux de la capitale fédérale. Le tablier, qui peut se soulever en son milieu pour laisser passer des bateaux, repose sur dix-huit piles en forme de « V »qui évoquent des mouettes en vol. De plus, les piles donnent à l'ouvrage une apparence légère et forment une arcade qui correspond à la forme traditionnelle des ponts de Washington. Une structure en acier a été coulée dans le béton afin de répartir la charge morte sur toutes les piles et de minimiser la poussée horizontale à laquelle sont soumises les fondations.

From left to right, from above to below:
View from shoreline, V-shaped
pylons, panorama view.
Right: Panorama view.

Von links nach rechts, von oben nach unten:
Blick vom Ufer, v-förmige
Pfeiler, Gesamtansicht.
Rechts: Gesamtansicht.

De gauche à droite, du haut vers le bas:
Le pont vu de la rive, gros plan sur
les piles en « V », vue panoramique.
À droite: Autre vue panoramique.

INDEX.

Imprint
The Deutsche Nationalbibliothek lists this publication in the Deutsche Nationalbibliografie; detailed bibliographical data are available on the internet at http://dnb.d-nb.de.

ISBN 978-3-03768-025-4

1st edition 2010

Editorial staff: Marek Heinel
Translation: Marek Heinel (German), Marcel Saché (French Version), Cosima Talhouni (English)
Text editing: Cosima Talhouni
Graphic concept: Michaela Prinz
Layout: Marek Heinel, Georgia van Uffelen, Natascha Saupe